Joseph Hippolyt Pulte

Woman's Medical Guide

Joseph Hippolyt Pulte

Woman's Medical Guide

ISBN/EAN: 9783337811785

Printed in Europe, USA, Canada, Australia, Japan

Cover: Foto ©berggeist007 / pixelio.de

More available books at **www.hansebooks.com**

WOMAN'S MEDICAL GUIDE;

CONTAINING ESSAYS ON THE

PHYSICAL, MORAL AND EDUCATIONAL

DEVELOPMENT OF FEMALES,

AND THE

HOMEOPATHIC TREATMENT

OF THEIR DISEASES IN ALL PERIODS OF LIFE,

TOGETHER WITH

Directions for the Remedial use of Water and Gymnastics

By J. H. PULTE, M.D.

PROFESSOR OF OBSTETRICS AND DISEASES OF WOMEN AND CHILDREN; EMERITUS PROFESSOR OF CLINICAL MEDICINE IN THE WESTERN COLLEGE OF HOMEOPATHY; AUTHOR OF "HOMEOPATHIC DOMESTIC PHYSICIAN," ETC.

FOURTH EDITION, REVISED.

CINCINNATI:
SARGENT, WILSON & HINKLE.
LONDON: JAMES EPPS, 170 PICCADILLY.
1863.

Entered according to Act of Congress, in the year 1853, by
J. H PULTE, M. D.,
In the Clerk's Office of the District Court of the United States for the District of Ohio.

PREFACE TO SECOND EDITION.

THE continued attention every where bestowed upon improvements in the education of youth, is one of the most cheering signs of modern progress in real civilization; right views on this subject seem to spread more rapidly than ever, and already we can perceive a change for the better in the health and vigor of our youth.

The author feels thankful for having been permitted to contribute his mite to the promotion of this great and beneficent object, and this the more so, as his efforts in this respect, seem to have met with a general and hearty approval of the profession, the press and the public.

He hopes that this second revised edition will meet with the same favor.

THE AUTHOR.

CINCINNATI, May, 1859.

PREFACE TO FIRST EDITION.

In offering this work to the Public, the Author deems it appropriate to state some of the reasons which prompted its publication.

The spread of Homœopathy throughout the country has been very great, beyond even the expectations of its most sanguine advocates, and is daily increasing. Thousands of families depend on its efficacy in the most dangerous diseases, such as Asiatic Cholera, Scarlet and Typhus Fever, etc.; and thousands of intelligent mothers consider it the greatest blessing which science has yet bestowed upon them. While they have heretofore received their information respecting matters of general interest and usefulness, physical education, hygiene, etc. from Allopathic writers, they now, since their conversion to Homœopathy, expect the Homœopathic practitioner to furnish them with similar instruction. This reasonable desire the practicing physician, burdened with toil and care, can satisfy only at a great sacrifice of time and breath, if he attempts at all to convey the requisite information personally to each one of his lady patrons. Popular treatises devised for this purpose, will accomplish the object in every respect more satisfactorily.

No such work, however, has, as yet, appeared in our Homœopathic literature, at least none especially adapted to the instruction of woman, as to her physical and moral education, her destiny, and the treatment of female diseases.

This want the author intended to supply, by discussing these important topics in a popular and lucid style. Whether or not he has succeeded in his task, is for the reader to decide: if not, then the failure has been one of judgment, not of motive. As to the latter, the Author is conscious of having spared neither time nor research in endeavoring to make the book indeed, what he intended it to be, a WOMAN'S MEDICAL GUIDE, forming as it were, a supplement to his *Homœopathic Domestic Physician.*

THE AUTHOR.

CINCINNATI, May, 1853.

CONTENTS.

INTRODUCTION, - - - - - - - - - - - 13–18

PART I.

WOMAN'S

PHYSICAL AND MORAL DEVELOPMENT;

HER SOCIAL POSITION AND DESTINY.

CHAPTER I.

WOMAN.

1.—PHYSICAL CHARACTER, - - - - - - - - 21–31

Anatomical difference between the two sexes, 21: Difference in the external appearance, internal organs and bony structure, 23: The female has a different sphere of action from the male, 24: Comparison of the female system with the male, 25: The destiny of either based on their physical constitutions, 26: Division of the duties of life, 27: Organs peculiar to the female, 28: Description of their form, position and function, 29: Explanation of the changes in woman's physical economy, 30.

2.—MORAL AND INTELLECTUAL CHARACTER, - - - - - 31–43

No superiority of either sex in mental or physical endowments, 31: Difference of character based on unalterable laws of nature, 32: Woman physiologically considered, 34: Phrenological analysis of the female mind, 35: Moral region. 36: Intellectual region, 37: Reflective and perceptive faculties, 39: Region of the sentiments—imagination, 40: Its importance as one of the faculties which render woman pre-eminently social, 41:

3.—DESTINY, - - - - - - - - - - - - - - - 43–52

Woman in every respect man's equal, 43 : Christianity the only true restorer of woman's rights, 45: Emancipation of woman, its true meaning and legitimate object, 46: Duty of legislation, 46: Her destiny, based upon her physical and moral peculiarities, 47 : Woman's true position in the light of the gospel 48· Her position in the family as wife, mother, sister or daughter, 49 : Duties and responsibilities of a christian mother, 50: Her position in society and the state, 52.

CHAPTER II.

GIRL.

ITS INFANCY, - - - - - - - - - - - - - - 53–60

Difference of action between the male and female infant, 53: Swelling of the breasts in female infants, 54: Hygienic rules, bathing, exercise, air, food, clothing, 55: Vaccination, 61.

ITS GIRLHOOD,-------------------------------- 61–84

Importance of a proper development and early education of the physical system, 62: Danger of the intellectual education if too early commenced, 62: Our present system of education is wrong—the physical must precede the intellectual, 62: Each has separate ends to accomplish, 63: What they are and how they can be reached, 64: Dress of a girl, 66: Exercise, dancing, gymnastics, 66: Danger resulting from infant schools, 72: Bad results from the practice of awarding preferments and premiums in schools, 74: How a system of education should be organized, 78: Music, vocal and instrumental, 80: Moral and religious training, 80: Boarding-schools inefficient and dangerous, 81.

CHAPTER III.

MAIDEN OR YOUNG LADY.

Changes in the system, 85: Menstruation, 86: signs of maidenhood, 87: Moral and physical changes, 88: Puberty, 89: Its causes, 90: Description of the internal organs of generation, 90: Nature and origin of menstruation, 92: Moral development of the maiden, 94: Education at home and abroad, 96: Gymnastic exercises, 97: Early marriages are injurious, 98: Their causes, 99: Their prevention, 100: Runaway matches, 100: Legitimate time of marriage, 102: Education never finished, 103: Different kinds of education, 106: Dancing, arts and sciences, 107: Extravagance in dress, 109: Duty of parents, 110: Show mania and fashions, 111: Necessity of useful occupations, 112: customs of old times, 116: Study of languages, 118: Moral and religious duties, 119.

CHAPTER. IV.

MAIDEN LADY.

Law of development, 120: Exceptions, 121: Duty to marry, 122: Duty and worth of maiden ladies, 124: Their occupations, 128: Their joys and pleasures, 130.

CHAPTER V.

MARRIED LADY.

Civilization of the world by woman, 132: Position of a wife, 133: Affection the real talisman of the marriage union, 135: Other virtues 137: Physical changes during marriage, 142: Conception, 143: Pregnancy, 144: Its signs, 145: The fœtus and its development, 147: Rules to be observed during pregnancy, 149: Its disorders, 150: Quickening—its meaning, 150: Duration of pregnancy, 152: Labor, 152: Chloroform and ether—their use, 153: Duties of a mother, 153: In the family, 155: In society, 158: In the state, 159.

CHAPTER VI.

WIDOW.

Widowhood, 161: External circumstances, 163: Duties of widows, 164: Second marriages, 165: Education of children, 166: Physical welfare, 167.

CHAPTER VII.

MATRON.

Her physical condition, 169: Her relations to family and society, 170: Her position and needs, 173: Recapitulation, 175–178.

PART II.

DISEASES OF WOMEN:

THEIR DESCRIPTION AND HOMŒOPATHIC TREATMENT.

CHAPTER I.

DISEASES OF SEXUAL DEVELOPMENT.

1. Puberty and its Abnormal Appearance, - - - - 181–205

Chlorosis—green sickness, 182: Menstruation and its abnormal appearance, 186: Tardy menstruation, 187: Suppressed menstruation, 190: Too copious menstruation—flooding, 193: Menstruation of too long duration 196: Too late and too scanty, 196: Deviation of menses, 196: Too difficult, 197: Painful, 198: Their cessation or change of life, 199: Abnormal erotic sentiment—nymphomania, 200: Absence of erotic sentiment, 204: Sterility, 204.

2. Pregnancy, - - - - - - - - - - - - - 206–240

Plethora—congestion—fever, 206: Hemorrhages, 209: Hemorrhoids—piles, 210: Varicose veins, 212: Swelling of the feet and lower limbs, 213: tooth-ache, 214: Salivation, 214: Derangement of appetite, 215: Nausea and vomiting, 217: Diarrhea, 219: Constipation, 221: Dyspepsia—heart-burn—acid stomach, 222: Difficulty of swallowing, 222: Spasmodic pain and cramps, 223: Colic pains, 224: Disury—strangury—ischury, 224: Incontinence of urine, 225: Jaundice (icterus,) 225: Pain in the right side, 227: Asthma—congestion of the lungs—palpitation of the heart—spitting of blood—pleurisy,

228: Hacking cough, 229: Vertigo—congestion of the head, 229: Headache—fainting—sleeplessness—depression of spirits, 230: Neuralgic pains, 230: Spasmodic laughter—crying—sneezing—yawning, 231: Puerperal convulsions (*Eclampsia gravidarum,*) 232: Miscarriage (abortion) 234

3. PARTURITION, - - - - - - - - - - - - - 240–273

Labor, 243: Natural and preternatural, 244: Protracted, 246: Sudden cessation of, 247: Spurious or false labor-pains, 248: Excessively painful labor, 250: The waters (child's water,) 251: Delivery, 253: Apparent death—asphyxia of the infant, 254: After-birth, 256: Hemorrhage—flooding, 258: After-pains, 260: Confinement, 261: Lochial discharge, 264: Suppression of, 266: Excessive and protracted, 267: Offensive, sanious, 267: Childbed fever, 268: Milk-leg (*Phlegmasia alba dolens,*) 270: Mania in childbed, 272.

4. NURSING, - - - - - - - - - - - - - - 273–291

Milk-fever, 279: Ague in the breast—gathered breast, 280: Deterioration of milk, 283: Suppressed secretion of milk, 287: Excessive secretion of milk, 288: Deficiency of milk, 289: Sore nipples, 290.

CHAPTER II.

DISEASES OF GENERATIVE ORGANS.

Imperforation of the Hymen, 292: Inflammation of external parts, 293: Wounds on the same, 294: Oedematous swelling of the labia, 294: Pruritus—itching of the private parts, 295: Diseases of the vagina, 297: Leucorrhea (*fluor albus,*) 298: Diseases of the uterus, 302: Prolapsus uteri (falling of the womb,) 303: Its retroversion and anteversion, 308: Inflammation of the womb, 309: Irritable uterus (rheumatism and neuralgia of the womb,) 310: Polypus of the uterus, 312: Scirrhus and cancer of the womb, 315: Ulceration of the womb, 316: Dropsy of the womb, 317: Inflammation of the ovaries, 319: Ovarian dropsy, 322: Diseases of the breasts, 322: Scirrhus and cancer of the breasts, 324.

CHAPTER III.

DISEASES OF NERVOUS FUNCTION.

HYSTERIA, - - - - - - - - - - - - - - - - 325–332

Hysteria or vapors was formerly a fashionable disease, 325: Has become quite obsolete at the present day, 326: Derivation and signification of the word hysteria, 326: Its chronic character, 327: Description of an hysterical paroxysm (*globus hystericus,*) 328: Resembling epileptic fits, 329: Its mental causes, excess of joy, fear fright, anger, grief, home-sickness, unhappy love, jealousy, mortification insult, contradiction, chagrin and indignation, 330: Hysterical constitution, its treatment, 331: Hysteria diminishes of late years in intensity and frequency—the probable causes of this singular phenomenon, 332.

NOTICE TO PHARMACEUTISTS.

The medicines prescribed in this work are generally found among those usually contained in the boxes accompanying the books for domestic practice; it has been the intention that their potency or attenuation should be the same as recommended in the Author's "Homœopathic Domestic Physician," viz: the medicines taken from the vegetable kingdom in the third attenuation, those from the mineral in the sixth. If separate medicine chests should be made to accompany this book, Pharmaceutists will please to put them up in accordance with the above rule; containing the following medicines:

Aconite, Arnica, Arsenicum, Belladonna, Bryonia, Calcarea carbon., Cantharides, Capsicum, Carbo vegetabilis, Chamomile, China, Cocculus, Coffea, Colocynth, Conium, Crocus, Cuprum, Drosera, Dulcamara, Ferrum, Graphites, Hepar sulph., Hyoscamus, Ignatia, Iodium, Ipecacuanha, Lachesis, Lycopodium, Mercurius (vivus,) Natrum mur., Nitric acid, Nux vomica, Opium, Phosphorus, Phosphoric acid, Platina, Pulsatilla, Rheum, Rhus toxicodendron, Sabina, Secale, Sepia, Silicea. Spigelia, Stannum, Staphysagria, Stramonium, Sulphur, Tabacum, Tartar emetic, Thuja, Veratrum album; also the tinctures named in the Author's work on Domestic Practice.

INTRODUCTION.

It would be almost superfluous, at the present day, to preface a treatise like this with an apology for its appearance. Society in modern times, and particularly in our country, has assumed a decided direction towards the diffusion of knowledge amongst all classes and ages; the rich and the poor, the mechanic, merchant and literary devotee; the child, young man and woman, parents and aged persons—*all* are provided for in the distribution of intellectual food. The presses are teeming with the various productions, suitable for popular education, diffusing a mass of knowledge, which, in time, will ameliorate the condition of society. The most abstruse sciences, heretofore strictly and carefully hidden from the eyes of the people at large, now make their appearance in crowded halls before popular audiences, being received with applause and eagerly absorbed, when brought before them in language appropriate and easily understood. Thus, Astronomy and Metaphysics have been successfully treated of in popular lectures; natural sciences, in all their various phases, are familiar and pleasant visitors at the gatherings of the people. It may be truly said, our lecture-rooms continue to do, what the school-rooms have had no time to finish; and an intelligent, well-educated people, like ours, will pursue its studies, in spite of old-fashioned barriers and scientific cliques; having once imbibed the thirst for knowledge, it cannot be restrained in continuing to satisfy its cravings.

Another question, however, might be raised, as regards the *propriety* of treating in books of subjects, so delicate and private, that the modesty of the female sex naturally would shrink from their perusal. But it is an old adage, and a very true one: *To the pure all things are pure;* where the heart does not already yearn after the evil, and the imagination is not already perverted, finding her sole delight in the contemplation of impure pictures, a truthful and severe exposition of the laws governing the female system, cannot intoxicate the senses or degrade the moral taste; on the contrary, must ennoble the impulses of the heart by increasing the knowledge of laws and destinies in connection with the immense responsibilities, thus given to the choice of each one's judgment. Beside, we are convinced that it is possible to treat of the laws of nature within the limits of *perfect decorum;* and this without suppressing the truth or becoming unintelligible.

No one will deny the propriety of giving to woman all the instruction needed for herself and offspring, and as this knowledge has to be presented to her in some way, that which least offends her finer feelings will be considered the best. Woman is naturally timid, and refrains as long as possible from making inquiries and asking advice from male persons about her own health or that of her daughters, where the subject is a delicate one. And yet she must seek for information, or else irreparable damage might be done. She fears exposure, if it be only in conversation; her nature instinctively revolts against it. Here it is that a book, containing all the information she wants, frequently comes to her as a great relief; she can receive instruction through it, without exposing her needs to the ear of a male person, be he ever so well known to her. This reluctance and fear of exposure is so deeply rooted in females, that they frequently rather seek advice, if absolutely needed, from the physician, who is a stranger to them, than from their own family-physician. How much easier is it to consult the pages of a book, which written for their especial benefit, will inform them privately about subjects, on which they hesitate to converse in the presence of others.

Again, is it not all important that woman, the mother and guardian of our infants and children, should possess all the knowledge possible as to their rational training and education. An ignorant mother will have an excuse for the neglect which her offspring has to suffer; she can say it is not *her* fault; if she knew more about the education of children, she would not allow them to fall under the supervision of nurses, or persons still more incompetent. Give to the mother the requisite knowledge in *this* respect, and thousands of evils will be corrected, under which at present our infant world has to suffer. Inform the mother thoroughly as regards the physical and moral training of the young girl, and soon society will feel the blessing of such beneficent undertaking; the next generation will already tell of the difference. Instead of sickly and nervous women, whose sole duty seems to consist in cultivating fashionable life, with its soul and body-destroying consequences, you will behold strong and vigorous bodies and enlightened and sprightly minds, whose duty will not be to run constantly after pleasure or external ornaments, but who would rather want to be themselves a pleasure in the family-circle, and an ornament to society. Our present system of female education, its workings in the school and at home, are entirely wrong and deleterious for the physical and moral welfare of the daughters of the land, who soon will have to be its mothers. But the evil does not stop here; it increases in a fearful ratio, as generation follows generation, until society and the state itself is brought to the verge of ruin and destruction. Well may the philanthropist shudder at the prospect in the future, on beholding the present mode of female education among the wealthier classes. From over-tasking the young mind with studies for hours and days, interrupting thereby the growth of the physical system, the inevitable result is a morbid development of the nervous system, a damage, which in most cases is irreparable. An undue degree of ambition is the cause of this growing evil. Each mother wants to see *her* daughter excel above all others, in what? In strength, health, good sound sense, modest behavior, sprightly, cultivated mind, knowledge of house and social duties? No, not in these qualities; no, but in a display of mental and bodily fineries, called fashionable accomplishments.

Ambition is a noble quality of the soul, when followed within the limits of reason; but it becomes a scourge and a destroyer of life and happiness, if immoderately indulged in. In the case of education it becomes more than that; it becomes a crime. How many a naturally strong constitution has been thereby enfeebled for life, and how many a naturally feeble one has been made thereby to depart this life. These are weighty considerations which, we hope, will receive all the attention from those, for whose benefit they are here introduced.

Let each mother reflect well and deeply, before she allows her daughter to be thrown into the whirlpool of modern education; let her consider first, whether the child is physically strong enough, to go through with it; because it is indeed *hard* labor, to perform all that modern education proposes to have done. In the following pages we hope to convince parents, that substantial wrong is done to their children by too early and closely pursued studies; that the body must be allowed to develop itself first, before the mind is taxed so severely, as is the case at present. This relates particularly to the education of girls.

Our plan will be, to present first, in a general review, the character and destiny of woman—physically, morally and socially; after which we intend to contemplate woman in her different stages of development, as a girl, wife, mother and matron. In each of these periods she will be subject to physical changes, and have to perform different duties. It shall be our especial duty to give a clear and faithful picture of these different periods of development and to show how woman ought to be in each of them, in order to deal out and receive the greatest amount of real happiness in this world, in which she decidedly is the most beautiful as well as the noblest of sojourners.

PART I.

WOMAN'S

PHYSICAL AND MORAL DEVELOPMENT;

HER SOCIAL POSITION AND DESTINY.

CHAPTER I.—WOMAN GENERALLY CONSIDERED.
            1. PHYSICAL CHARACTER.
            2. MORAL AND INTELLECTUAL CHARACTER.
            3. DESTINY.
    "   II.—GIRL.
    "   III.—MAIDEN OR YOUNG LADY.
    "   IV.—MAIDEN-LADY.
    "   V.—MARRIED LADY.
    "   VI.—WIDOW.
    "   VII.—MATRON.

CHAPTER I.

WOMAN.

1. PHYSICAL CHARACTER.

In treating on this subject, we pre-suppose that the reader is familiar with the general outlines of human anatomy, and even with such details as have reference to the principal organs and their functions, necessary for the maintenance of life. These the female system shares in common with the male. It is our object here to point out only that, wherein they differ; and moreover to show that this difference in their physical nature is the cause of their respective destinies and duties.

In comparing the female with the male body, we find at once the former less in stature and weight, but more rounded and graceful in form. It exhibits not so much strength of the muscles; these being less developed than in the male. But its motions, if less agile and strong, evince more graceful elasticity; in this respect, as also in the weaker but clearer voice, the female has retained more or less the character of the child. We shall have occasion afterwards to remind the reader again on the many analogies between the female constitution and that of the child in general. This is mentioned here at once, in order to draw the attention to a fact, which will have great influence in the better understanding and appreciation of the female character and destiny.

Thus in the outer appearance a great difference is noticed between the two sexes, and this increases as we proceed to a closer examination of the several parts of the system. In doing so, we have to compare the female organization with the male.

The female face is smaller, its front not so high, but the neck is longer; the chest is not so capacious, but the abdominal region more developed particularly across the hips, which thus forms in the female the basis-line of a triangle, whose apex rests in the region of the shoulders; while in the male system the order is reversed, the shoulders forming the basis-line and the hips the apex of the triangle. The female has shorter, but rounder and fatter arms, with softer contours; the hand also is smaller, whiter and softer, the fingers are finer and more pointed. The lower limbs, on the contrary, although shorter, are stouter, particularly from above downwards to the inner corners of the knee, which are generally rounder; the feet are shorter and smaller. The female skin is finer, softer and more transparent, the veins are easier discernable; on the more delicate parts of the skin, such as the face and neck, the blush appears quicker and easier. The hair grows more abundantly and longer on the head, but less so on other parts of the body; the hair itself is finer, softer, more elastic and glossy, the nails are more transparent and tender. The areolar tissue, which contains the fat, is more abundant and firmer; the muscles display a lighter color, are softer, thinner and weaker; the single muscles less protruding. The diaphragm is less in size and lies higher up, enlarging thereby the abdominal cavity; the heart is smaller; the walls of the whole vascular system are thinner; particularly does the arterial system lack the same density of texture in proportion to the venous, as this is the case in the male, although the female possesses a larger amount of blood in proportion. The lungs are smaller, and the apertures of the nose and the whole wind-pipe narrower. The brain, in proportion to the body, is heavier in the female, than in the male; the cranial cavity is more elevated as regards the position of the face, and the mass of the brain in proportion to the nervous system and the cranial arteries is larger than in the male. The nerves themselves, however, are finer, more tender and weaker.

The bones are thinner, smoother, less firm, their processes and indentations less distinct; there is less bony matter in general; the bones of the face are finer, with a more even surface; the cavities in the front part of the head and cheeks are narrower; the ribs are thinner, flatter, shorter, starting in a more decided curve from the spine; the spinal column therefore reaches deeper in the cavity of the chest, and the spinal processes are less discernable in the back-bone; the breast-bone is shorter; the cavity of the chest generally less capacious; the spinal column, as a whole, is rather longer in the female. But the greatest differences exist in the region of the hips, the bones of which are broader, although they are thinner and lighter: the cavity of the pelvis inside, formed by the hip-bones and the small of the back, is every way longer, wider and more uniformly so, above as well as below.

We must be convinced by the above comparison, of the female system with the male, that the former was destined to move in a different sphere of action from the latter, and that this cannot be changed without serious danger for its physical welfare. The duties and mode of life seem to be clearly pointed out to the female in her bodily construction, which we now will proceed to consider, before we come to describe those organs, peculiar to the female sex, in which the greatest part of her destiny is fixed.

As a prominent difference, we find the framework of her system less in size and strength; the muscles, attaching to it, are also smaller and weaker. This fact shows that the female body is less capable than the male of undergoing toil and hardships, which require mostly muscular effort. Neither would it be possible to make her equal in that respect to the male, because her muscles are softer, thinner, not so compact and of lighter color; this physiological difference precludes them from ever becoming so compact, hard and strong, as the muscles of a man, even if ever so much exercised for that purpose. Besides, it states that the female has a firmer areolar tissue, and in greater abundance than the male; consequently is more inclined to become fat, and her form rounded, a circumstance which prevents any extraordinary degree of muscular development by exercise. Yet this latter is on that account not the less needed; enough of it must be taken by the female to develop those changes within her system, which, as we will see hereafter, are a peculiar characteristic of her sex.

Another marked difference consists in the form and capacity of the chest and abdominal cavity, compared with that of the male. The cavity of the chest is smaller, while that of the abdomen is larger. Here we find a wonderful and wise provision made for the respective duties of the two sexes. The male is destined to labor harder; to him is given the large wide chest, attached to which are the powerful muscles of the trunk and arms; his lungs are of greater expanse, his arterial system, including the heart, is more developed; all this is needed for a more vigorous and prolonged prosecution of his daily manual work; his system is fitted out for that especial purpose; the abdominal region is less in size, but nevertheless strong and compact, while his lower limbs principally excel in elasticity and propelling power. How beautifully is his destiny written in the framework of his body and the display of his muscles. On the other hand, the female is larger across the hips, because there nature needed above all other parts room to fitly prepare woman for her destiny. From this is easily seen, how perverse to nature that fashion is, which compresses the female waist and makes of her literally an hour-glass, or a wasp, while, indeed, her form ought to be broad across the hips, gradually, but regularly tapering off upwards; the male form being just the reverse of this.

The nervous system and the brain of the female show another great difference, indicative of her duty and destiny. Her brain is proportionally larger than that of the man, in comparison to the size of the body; but her nerves are finer, more tender and weaker, making them disproportionate to the size of the brain. The nerves, which are the carriers of the brain-power, ought to be developed, in proportion to the size of the brain, in order to fulfil its demands; they are, however, finer and weaker in the female, hence she lacks power of execution, which compels her to look for other help to have her designs executed. By this arrangement she has become pre-eminently the counsellor and constant companion of man, who is well fitted by nature to carry out her designs, together with his own. This is one of the most wonderful arrangements in nature, by which the two sexes are bound indissolubly together, as the intellectual intercourse between them is rendered thereby a natural necessity. *He* flies to her in time of need for counsel and advice, and *she* looks to him for action and execution. We can here only slightly touch upon this interesting condition of the two sexes. Hereafter, however, we will treat of it more fully.

Thus, we see, that the domain of action for the female is at the *side* of man, not ahead of him, nor in his rear. These two have to accomplish *together*, what it would be impossible for one to perform alone. The hardships of life's duties are thus wisely divided, while the results from their faithful performance are mutually enjoyed, each one receiving a *double* share of benefits for *one* share of labor. The female, physically disabled from participating in life's hardest struggles, receives a higher trust in the composition of her moral faculties, by which she is enabled to buoy up the toiling partner, when he seems almost to sink under the too heavy burden.

Her sphere of action will be always more or less the house, with its manifold labors and attractions; there the daughter, wife or mother finds room and time enough to exercise her physical and moral powers; and if no morbid craving has taken possession of her mind, she will find real contentment and happiness. Beside, the social circle, the school, lecture-room and church offer sufficient opportunities for the display of her intellectual faculties, in giving or receiving instruction. If inclination or necessity prompts her to devote herself to duties out of the house, society presents thousands of occupations, which she can fill honorably and profitably, without risking her health or exposing her feelings. We will now proceed to call the reader's attention to those physical peculiarities, which belong to woman exclusively, and form her sexual character, determining thereby more than by any thing else her true destiny.

It would be impossible to give an anatomical description of all the organs peculiar to the female system, without illustrations; yet some definite idea must be had about their position and structure, in order to show their harmonious co-operation and complex relation to each other. We prefer, however, not to introduce into this work illustrations, which might be considered objectionable on the score of propriety and good taste. To obviate this difficulty, therefore, we will compare the position, form and size of these organs to familiarly known objects. This glance will suffice to give to the reader a sufficient knowledge of these parts, and if any farther should be desired, every one can easily have access to anatomical works, treating fully on this subject. The region between the hips contains within itself a cavity, called the pelvic, and the bones, which form this cavity, the *pelvis;* it means literally a bowl or basin. Its wider margin is above, its narrow opening below. The pelvis contains within its cavity all the organs we intend to notice here.

The most important of these organs is the uterus, or womb, which occupies the centre of this cavity, suspended on ligaments on either side, and connected by very loose tissue in the front with the bladder, and in the rear with the rectum, the natural outlet of the contents of the bowels. The uterus has a pear-shape, and is about as large as a small egg, with an oblong opening of one-eighth of an inch long, across its smaller end, called the mouth of the uterus, which leads into its cavity. This cavity in the unfecundated uterus, is very small, and only sufficient to be noticed as such from the mouth upwards to its higher and thicker part, called the fundus uteri; here, to the right and left, are two small openings, which connect through a tube on each side, called the Fallopian tube, with the ovaries, two oblong, flattened and oval bodies, of a whitish color, and the size of half a walnut. They are called ovaries, literally egg-beds, because they contain the ovum or egg, which is formed there, and when growing, bursts the outer skin of the ovarium, whence it is transferred through the Fallopian tube into the uterus, where in single life it passes away with the menstrual discharge; in married life however, when all the conditions of nature have been fulfilled, is retained within the uterus, and nourished there during nine months. While gestation is progressing, no new developments of eggs in the ovaries take place; at least that is the rule; but very few exceptions occur to the contrary. Generally speaking, the action of the ovaries is suspended during the time of gestation.

This interesting period in a woman's life brings about a number of very important changes in her physical economy.

The uterus, a very small organ before, now has increased to a great size, and its former insignificant cavity has become immensely enlarged. Organs which formerly were active, now lay dormant, and other organs, formerly asleep, now have become operative. Menstruation, the regular monthly discharge, has ceased, and the fluids, formerly thus wasted, now have become the source of life and nourishment for a new being, and before this is yet born, the breasts begin to swell, preparing to yield the same precious nourishment in another more appropriate form and place to the child after its birth. These are some of the changes in the female system during this time, and they are greater in extent and importance than any other system has to experience in its life time. No wonder that they are often accompanied with great danger, or prepare derangements of health, which often last for years. We will have occasion, when speaking of the diseases of woman, to trace a number of them back to this period of her existence, where either by neglect or accident, their seeds were sown.

2. MORAL AND INTELLECTUAL CHARACTER.

We have seen that for wise purposes the Creator endowed woman with a physical constitution different, in many respects, from that of man; each one being deficient and lacking perfection in several particulars, but when united in the sacred bonds of marriage and laboring together to build up a family, these two present an exalted picture of perfection, of which nature has nothing equal to show. The strength of the man is united to the loveliness and grace of woman; his impetuous but noble haste is checked by her timid and cautious foresight; his bold aggressive spirit tempered by her wise reluctance of action; everywhere, a blending of qualities, which hides each other's faults, while it admits their virtues to appear the more readily. The same wonderful arrangement can be discerned throughout the moral constitution of the two sexes, each of which, taken separately, may exhibit many faults and weaknesses, which mostly disappear when coming in contact with the good qualities of the other. From this point of view we have a clear insight into the wonderful operations of Divine Wisdom, which ordered just such relations as we behold, to exist between man and woman, endowing each one with different qualities, to promote the happiness of both.

From this, it is evident that, strictly speaking, we cannot grant to either sex a superiority in gifts or qualities, either physical or mental. Each one stands on a separate platform, distinct from the other, and there appears unrivalled in some respects while deficient in others. Yet both have received a sufficient share of those high qualities which stamp the image of God on the human form and soul, and none is farther away from this ideal of perfection than the one who would be low enough not to recognize this divine inheritance in the other to its fullest extent.

It is part of the Divine economy to have thus endowed the two sexes differently, in order to insure a more perfect union between them; they find in each other enough to admire and to love, as also from each other enough to learn and to imitate. In the likeness, but not the sameness of the two sexes, consists the most perfect harmony. They would not find pleasure in each others company, if their intellectual and moral composition were identical; the spice of intercourse, the charm of novelty, would be wanting; no inducement to progress, so essential to the welfare of intelligence, would exist; no social feeling would bind families together and create State organizations, because the family hearth, with its domestic happiness as the foundation, would be wanting. From whatever view we may contemplate the relations between the two sexes, we come to the firm conclusion, that in every respect they are equals in position, although not identical in physical or mental composition. The first condition insures to each an equal share of prerogative and emolument, while the latter fixes for both their respective duties and labors. Let us here yet remark, that revelation as well as physical and moral laws corroborate the above defined relations between man and woman, and if state and society had carried them out fully, untold agonies of body and soul, innumerable crimes and heart-rending scenes might have been averted, and the history of mankind been made to flow like the rivers of Paradise, full with the waters of joyous life, instead of what has been the case, full of blood and destruction.

The phrase, "*emancipation of woman*," should never have been thought of; its very sound is a disgrace to language, the especial gift of the Creator. And it never would have been uttered by suffering woman if she had received ample justice from man as regards her social and political rights, particularly those of property and labor. Of this, however, more anon.

Generally speaking, we can be justified in asserting, that *moral* sentiments are more largely diffused among women than men; their veneration and benevolence are largely developed and make them particularly well qualified to perform their angelic mission upon earth. They also have received a greater share of hope and conscientiousness, which buoy them up in the most disagreeable situations of life, and secure to them the peace of mind and the charity so characteristic of the female sex. These four faculties comprise the whole of moral sentiment, and are those by which man is particularly distinguished from the animal. They make him sympathise with others in misfortune, and love and treat with kindness and humanity, the poor, aged and infirm. They link his spirit to the Deity in adoration and love, while he is made to submit cheerfully and easily to the Divine will under all circumstances. Such exalted sentiments have been given to woman more abundantly than to man, and make her, in the sight of the latter, an object of veneration, regard and love, even if no other superior quality should adorn her mind or grace her figure. This is so universally true, that man, even in the most barbarous periods of history, never has failed to show this almost religious veneration to womanhood, which fact is verified by many instances amongst the earliest nations. A kind-hearted, benevolent, pious woman, will always be an object of general admiration and reverence, while the opposite character will meet with as universal neglect and disrespect. In endowing woman so freely with these high and ennobling qualities, the Creator threw around her feeble frame and position, a far more potent shield to protect her, than any other instrumentality could have afforded. Womanhood never appears to better advantage, than in the holy garments of moral purity and divine consciousness. Then she is irresistible and all-powerful; there she seems to be all at home; no gift of the intellect, no dazzling wit and splendor of beauty, can compensate for the want of such exalted excellence.

Those who seek to locate the power of woman in her superior external beauty, and rest her claims on these frail pillars, do not understand her proper relation to the other sex, and fail, entirely, to comprehend the *true* strength of her position. A beautiful face and graceful form will shield her often from the attacks of the low and rude, but never, on that account alone, secure for her the esteem and reverence of the noble and refined. These she can secure only by the excellence of her moral character, of which she has received so large a share. It is, therefore, part of her earthly mission, to let the power of these high qualities be felt as much as possible, by teaching their principles to the young, and acting according to them before man. If she does not do so, she is doubly guilty of neglect of duty and propriety, since she has treated with contempt the greatest gift nature has so freely bestowed upon her. It is far easier for her than him to cultivate the strictest morals, because she inclines to them almost by instinct. Her trials and temptations in this sphere of life's actions, are less severe than those of man, and consequently her distinctions in this respect far less meritorious. If, therefore, she leads man on the moral path, she ought to do it with that modesty of behavior which does not let him feel her own superiority, else one-half the benefit might be lost thereby. She ought to remember that his share of the moral faculties is less than hers, that she feels, by intuition, what is right and proper, while he has to try to arrive at the same point of right feeling and acting, by reflection and reasoning, a process, slower, but more tenacious, and when successful, even more exalted in its results. Thus, this apparently great disparity between the sexes, has been made the source of the greatest blessings to mankind, and if rightly understood and practiced, must render the world a paradise.

In the *intellectual* region woman has been endowed more with *perceptive* or *observing* than *retentive* faculties. She has a keen perception of all that passes around her, particularly when it has reference to herself; very seldom, however, does she reflect on the nature of the object she observes, or the probable effect it may have. She is satisfied with having noticed it without further speculation. This faculty gives her a proneness to curiosity, gossiping and light talk, which, if indulged in, must weaken, in a great measure, the influence which her nice discrimination in moral matters could otherwise secure to her. It gives rise to another fault, so frequently met with among women, that of searching for each other's failings, rather than virtues. If woman shows lack of intellect by the side of man, it is more in the deficient discrimination of the worth of others and its acknowledgment, than in any other respect. She has, generally, when called upon, just as clear and correct a judgment as man, and equal firmness and will, but less generosity and justice in the appreciation of others. And this is more apparent when she reviews one of her own sex, in which case her critical acumen becomes truly formidable and unmerciful. No fault, ever so slight, escapes her notice, while the good qualities of the person under notice, are generally overlooked. This trait in the female character, strangely contrasts with her otherwise benevolent tendency, and becomes, when not properly checked, a fruitful source of all kinds of annoyances and unhappy feelings, which often reflect even disadvantageously on her physical health. How many diseases have their exciting causes in nervous irritation? The nervous system of woman is easily affected, and the conditions of life in which women are frequently placed, render her still more sensitive to mental irritation. In child-bed, for instance, the slightest unkind word may produce the most disastrous consequences, and be destructive to both mother and child. How necessary it becomes, therefore, to cultivate kind feelings towards all with whom we have to associate, or about whom we have to express an opinion. A censorious spirit, if allowed to come up within us, soon grows to a height of intolerance, bigotry, and selfishness, which embitters the life of its owner forever. And woman, by her keener perception and observation of personal matters, has to fear its tyrannical sway more than man, who is generally more reserved in expressing opinions about others, even if his judgment in regard to them would be the same. Woman perceives quickly and expresses readily, much more frequently than prudence would allow. This is owing to a less intense action of the *reflective* faculties, causality and comparison. She can easily, however, remedy this seeming fault in her nature, by considering well before she expresses her thoughts, thus exercising her reflective faculties. Her innate benevolence, too, must be exercised to keep down gossiping and censoriousness, as it is a benevolent disposition, particularly, which constitutes the character of a true lady. Such an one will always be careful never to wound the feelings of another by words, gestures or otherwise, be the object of her remarks ever so insignificant. The golden rule contains for no one a more precious precept than for woman, whose conduct ought to be at all times measured by it.

Amongst the *perceptive* faculties we must mention, especially, order, of which woman has received, generally, a larger share than man. This organ is the soul and ornament of a well directed household. To its influence, the domestic hearth is mainly indebted for its charms, and civilization considers it one of its main-springs. Its exercise and cultivation, therefore, are of the highest importance to every woman, as part of her individual character, in producing and representing in the family circle, that divine principle of order which the Great Architect so scrupulously observes in the whole Universe.

In the region of the *sentiments*, woman is also richly endowed. She possesses, and ought to cultivate a proper self-respect, a feeling of womanly pride, so as to make her independent in thought and action, without rendering her haughty or presumptuous. She can and ought to gratify a desire to excel and please, of course in a moderate degree, lest it might degenerate into vanity. Her love of approbation is generally large, and while she has less caution than man, she is in greater danger of becoming vain and coquettish. Proper self-respect, however, will soon correct this evil tendency.

Her imagination is splendid and generally more brilliant and quicker than man's, and in connection with greater and readier humor, renders her social qualities far superior to those of the male sex. She possesses a quick and lively conception of the ridiculous, even in such a degree, that it frequently ought to be restrained. Without woman, society would be barren of interest. No mirth, no merriment, no pleasantry and wit would take away the tedium of intercourse. Mankind would have lost the elasticity of its step; and as these qualities are essentially preservative as regards life, it is evident that woman, in this respect contributes an equal share in the maintenance of social order with man, who so often prides himself on being the pillar of state and society, overlooking, in his ignorance or pride, the essential benefit and help he continually receives from the feebler sex in supporting him to bear his burdens. It is needed for woman to know the importance of her office and duties in this respect, as it will make her love and cultivate qualities so characteristic of her sex, and so essential for the welfare, even the physical, of mankind. Yes, truly may we say, the welfare of mankind depends, in a great measure, on these eminently social qualities of woman, infusing joyfulness, hilarity and buoyancy into every-day life, thereby lessening its burdens, promoting physical health and moral strength. Man forgets, in her pleasing company, the earnestness and severity of his thoughts and pursuits. He, who but shortly before was deeply engaged in the serious conflicts of life, finds himself disenthralled from such fetters, when addressed by her gay, lively and buoyant conversation; he is caught by her spirit in the pleasant retreat of imaginative sentiment and the enticing flow of humor, wit and conversational entertainment. For the welfare of his body and soul, this rest from fatiguing business is indispensable. If it were not so, his powers would soon be exhausted, and languor and disgust unfit him for the further performance of his duties. Physically and morally, therefore, these exhilarating faculties of woman, are essential to the welfare of the whole race; and for woman herself, their cultivation becomes of the highest importance, as she thereby perfects her natural gifts, and holds undisputed rule in those holier spheres of life, the social circles, where spirit communes with spirit for nobler purposes and enjoyments.

Closely allied to the social circle, and, indeed its very prototype and basis, is the family circle, the domestic hearth, which, without the presence of woman, would be desolate enough. Here, also, she reigns supreme. Her delicate feelings, amiability, filial and parental affection, make the home where she resides, truly a paradise for man, who, without it, would be the most miserable of mortals. Those qualities, so extremely predominant in the female sex, are still more active, because not disturbed much by the selfish propensities, such as love of gain, of which man has received so large a share.

In the above, we have tried to give the general outlines of what *should* constitute the moral and intellectual character of woman. To exhaust the subject fully here, is impossible, for want of space, as it would take a volume alone to do it justice. Our object is to draw the attention to those characteristics of the female sex which fix her earthly destiny, and which guide us in the selection of the best methods for her physical and moral education, from her earliest infancy up. This subject ought to receive our most careful attention, as, according to our opinion, the welfare of society depends, mostly, on the soundness of the female sex, physically, intellectually and morally.

3. DESTINY.

From the preceding, the reader will perceive that, in as far as woman is different from man in her physical and moral constitution, her destiny cannot be identical with his; that while the strength of his physical frame points out a sphere of action for him, which is filled with hardships of all kind, her more delicate body must naturally be suitable and inclining only for tasks less severe. This is also true as regards intellectual labors. Their correspondence in this respect is perfect, although it might be said, that woman had heretofore, and was yet engaged in mental and physical transactions, as arduous or weighty as ever had fallen to the lot of man. This certainly is so, but it can only be a farther proof of the truth, that exceptions even in this highest productive sphere of nature, the creation of man, confirm a rule. As exceptions, we easily understand and value the acts of those who have played, during their life-time, the part assigned to the opposite sex. Such occurrences are frequently met with in history and daily life, and demonstrate the identity of the human forces, and the harmony of their tendency. They show the faintness of the line of distinction, where the two sexes meet, and the ease with which parties of either side may overleap it. We have feminine men, as well as masculine women. But these cases do not furnish a rule. Nature has drawn a line, and its existence can only be doubted by those whose inclinations have carried them already beyond its limits. The workings of society have shown this already for thousands of years, during which the duties and affairs of both sexes have been more or less clearly defined. It is true certain ages have not done justice to the claims of woman, rendering her lot harder than it should have been. But these times were yet barbarous and savage. The light of the gospel had not penetrated their darkness, and physical force was their only law-giver. How could woman, with her inferior physical strength, be treated on terms of equality by man, who measured everything around him by the strength of his arm and the force of his blow. If nature had designed woman to be equal in physical power, why did it not manifest itself in these times of physical preponderance? why did woman not then assume the place occupied by man in society? why did she not fight the battles and rove about, bent upon plunder and robbery? Why did she submit to a treatment as unjust as it was cruel? That she however did submit, and silently suffered for ages, is an historical fact, and proves evidently woman's superiority of moral strength in enduring the wrongs inflicted upon her by man's undeveloped intellect and moral faculties.

But the attitude of the two sexes changed as soon as the light of the gospel penetrated the spiritual darkness of the world. Christianity restored to woman her rights, and put her in the only true position by the *side* of man, where she always should have been, but never had been before, and never will be except when brought and sustained there by the doctrines of the Saviour. One of the last and most tender and affecting acts of His life had reference to this very relation of man to woman. He charged one of his disciples to take care of His mother; "and," saith John, "from that hour that disciple took her unto his own home." Here was a relation established by the Holy One Himself, between two persons whom He loved, and thus, according to nature's law, wanted to provide for. The woman was protected and taken care of by the man, the feeble by the strong, who in return received a mother's love. Thus it always should be. To destroy this relation of man to woman would be cruel, because tearing asunder the tenderest chords of human sympathy, based upon mutual dependence, and would be sacreligious, because laying violent hands upon the laws imposed on nature by the Creator. Views different from the above, have lately arisen in various parts of the world, claiming for woman equality in *all* things with man; they call for an emancipation of womanhood, as they term it. The endeavors of these reformers may be well grounded as to certain evils yet existing, such as the unequal right to property between husband and wife.* They can, however, not be justified as to the extent of changes which these reformers are aiming to accomplish. To make woman participate in all the gross and inglorious, but necessary work of man, would destroy the true glory of woman's existence, annul her dignity, and poison the innocence of her heart with desires entirely foreign to her nature. What a sight to see the graceful form of woman mingle in a political crowd, eager to deposit her vote or to share in popular distinction.

---

*We have no doubt that the progressive intelligence of legislation will remove, in time, all the civil inconveniences to which woman at present may be subject. One by one these impediments will have to yield before the bold and liberal reasonings of the age. As regards the question of property between the two sexes, we venture to predict, that one day it will be found just and right to allow the woman to be the owner in fact of one-half of all the property that is acquired during the marriage term; that she also will be the equal loser in all transactions to which she, either in writing or orally, had given her assent but not to be a loser where she thus had not assented. If she would lend her husband her money or other property, she should have the same right against him which other creditors have; that, however, her personal property, such as jewelry, should not be exempted from being taken to pay either her own or the debts of the matrimonial firm; but should be exempt in case where the husband contracted the debt without her consent.

It takes the massive soul of man to become reconciled to duties and honors of this kind, which he must be firmly persuaded to owe to his country, lest he would find them too ardous and insufficient. Again, how revolting the spectacle, to see women engaged in occupations which would not suit the delicacy of their form and feelings. Hard, severe bodily labor would destroy their beauty and be ruinous to their health. What would become of nome, its duties and pleasures? Who would have them to perform and enjoy? Can there social harmony exist, to charm and bless life, where no diverse directions of pursuit will allow of a uniting angle? All would be disunion, because one direction of pursuit would keep the pursuers in parallel lines, which never unite.

We could thus continue to bring arguments without number against the so-called defenders of woman's rights, if it were necessary to do so. The destiny of woman is so evidently distinct from that of man, and indicated so clearly in her physical and moral construction, that we can safely leave the settlement of this question with the sound judgment of the reader.

We will now proceed further to define the true position, which woman ought to fill here on earth according to her physical and moral character. These latter, as we stated above, have only received their due weight in the social order, since the gospel of love appeared among us, to supercede the gospel of law. Nothing is clearer proved by past history and the present condition of heathen communities, wherein woman is still treated far beneath her high moral endowments. And this close sympathy between the religion of Christ and the true social condition of woman, is still more apparent when we consider the incontestable fact, that as Christianity appears in a country more true and active, in that same degree will its women be more elevated honored and loved. It is, therefore, the Christian woman, in her relation as wife, mother and daughter, whose duties and destiny we here want to lay before the reader.

"It is not good that man should be alone; I will make an help meet for him," was the simple, but grand and effective speech which preceded the creation of woman. Society should be established; it had become a necessity, and woman was needed, just such an one as would fulfill the demands, thus made on her by the Creator. The laws of her life should be in harmony with these demands, that she might be indeed "an help meet for him." And, truly, it is so yet; the divine designs can yet be traced, sometimes, even in perfect purity, in the relations between the two sexes. Although the fall of man has rendered the original types less defined, and it is but too true that the first paradise is lost, still the *Christian* woman is able to regain it, partially at least, for man, if the latter is able to appreciate it. Such are yet her lovely endowments bestowed upon her by the Creator, for establishing society, and such is yet the longing of man towards her company, her consoling, inspiring presence and the sweet interchange of sentiments and ideas.

Woman is destined, therefore, to create and rule society; she was created for that purpose, her duties point that way. The family, with its various small but continual cares, falls to her superintendance. Her watching eye and intelligent rule must be felt in every corner of a well regulated household. Being the mother of the children, they continue to cling to her far beyond earliest infancy. She has almost the sole control of them up to the time when the permanent teeth appear, after which she divides the care over them with husband and teacher. This, however, does not diminish the importance of her part in the education of the child more advanced in age. Besides its physical welfare, the care of which she retains to the last, her principal duty will then be to cultivate, by example and teaching, those highest of all faculties, the moral, which, in their bearing upon temporal and eternal happiness, are far superior to the mere intellectual, the cultivation of which, falls to the lot of man. Thus she will be principally the one who teaches the child to be conscientious, modest and benevolent. Her influence will be mostly instrumental in leading the dawning sparks of veneration into the culminating centre of religion, from which radiate the all-inspiring rays of hope and eternal blessedness. No other agency in the place of a faithful christian mother, can accomplish this highest of ends so well. She should not, under any consideration, relinquish *this* part of her duty, as it involves the dearest interests here and hereafter. How important, therefore, is it for her to appreciate, fully, this great responsibility of her existence, and to know well how to fulfill all its demands correctly. An intelligent, pious mother, will look around to gain all the necessary information on this point, lest she might err in judgment and execution. She feels her duty and is not unmindful of the high reward which awaits her in beholding the prosperity of her children. She knows it was, under God, her own work. But who will describe the pangs of a mother's soul, when the object of her love turns out to be an outcast from society, and an object hated by God and man? If she was conscious of having neglected her part in his education, how fearfully must increase that remorse and agony! "The *immense responsibility* of parents cannot be too highly estimated," says Mrs. C. M. Steele.* "On it

---

* We recommend to the particular attention of our readers, a little work of Mrs. C. M. Steele entitled: "A Mother's Thoughts on Parental Responsibility."

rests the beauty and loveliness of the structure of mind? Unless *mothers*, who have the first nurturing of these delicate gems, are *fully* sensible of what awful results will flow from their hands in the neglect of the trust reposed in them, they must be unfitted for their work. If disregarded, what a dreadful loss must necessarily and inevitably follow. O, that the most vivid consciousness of this truth might irradiate every benighted female intellect, that she may never be compelled to perceive the direful images of this neglect."

Another part of woman's destiny is to be an help meet for man. She is a wife, sister or daughter. In either of these conditions, the variety of duties does not change the object of her life. They all point, from their various directions, to the one great purpose to be accomplished by woman, as his companion and help meet. She shares his griefs and sorrows, lessening their severity by quieting the storm of his heaving soul, and by elevating the hope and courage of his drooping mind. She partakes of his joys and pleasures, sympathising with his elated heart, but purifying its outburst by the refinement of her taste and the chastity of her feelings. She is his guardian angel in times of temptation; her advice and enthusiasm encourage and support him in times of peril. She becomes his *alter ego*, the better and purer principle of his own self. Without her, his loneliness would be insufferable; his misery complete.

Such is woman, the companion and help meet of man. This is her destiny and duty in the family circle, which she creates and sustains with her natural gifts, adapted to that purpose.

The same faculties which make woman the ruling and beloved mistress of the house, lead her into society at large, of which she is the soul and idol. She contributes more than man to the establishment of mutual friendship and its social exchange. Her heart swells easier with humanitary feelings. She is affectionate, and being less selfish and retired, willing to know and inquire into the conditions of others. Hence her strong social tendency, which is one of the greatest civilizing principles, to which mankind is mainly indebted for its progress. How important is it, therefore, for woman to cultivate her social gifts, in the right direction, in order to fulfill, satisfactorily, this, her glorious mission!

CHAPTER II.

GIRL.

ITS INFANCY.

In earliest infancy, the difference of action between the male and female is hardly great enough to make it an object of particular notice. Yet, it soon becomes preceptible to one, who takes the pains to observe closely. The infant female has generally a more delicate look; its frame is less massive, the limbs more slender, and the traces in the face finer and sharper. Its nervous system can be agitated more quickly, it is more susceptible of external impressions, and their re-action on her internal preception is easier but less enduring. A female child must, therefore, be more liable to disease, and less able to withstand its attacks. But it might not become, for that very reason, as seriously sick as the male. Its impressions are not so deep and lasting. Nevertheless, we know that the greater average mortality of children is on the side of the female portion. It is true, a prominent cause of this phenomenon, may lie in the fact, that in general more females are born than males. This, however, could not account altogether for the greater mortality among female children; we must find a part of its origin in the feebler physical constitution of the female child. It is, therefore, our duty to be more careful in nursing and training the female infant, than we have been heretofore, in order to effectually remedy this evil. The child's physical constitution must guide us in the selection of these rules. We dare not follow the old routine, merely because it is sacred by age, or agreeable to custom or fashion. Old usage, custom and fashion might kill our darling. Let us by all means save its life and preserve or restore its health, the most precious gift on earth.

We presume the reader is acquainted with all that pertains to the good nursing of children, generally; if not, information can be had in books called "Homœopathic Domestic Physicians," which treat fully on the principles of nursing.* We must at present, confine our remarks on this subject to those peculiarities which, in the bringing up of a female child, have to be especially attended to. These are, it is true, but very few; but not altogether unimportant, as will be seen directly.

---

*See in my "*Homœopathic Domestic Physician,*" the article, *Treatment of Children.*

During the first weeks after the birth of a child, there occurs sometimes a swelling of the external breasts, which, though it will be of little consequence to the future welfare of a male infant, frequently destroys forever, that of a female. This inflammation, if badly treated, may terminate in induration or atrophy of the milk glands, and thereby deprive the future mother of the use of these most important organs. Thus not alone she, but also her offspring will have to suffer severely for want of a little more than ordinary care in cases of this kind. Mothers and nurses ought to be well instructed in regard to the treatment of this disease; they will find the necessary information in the second part of this work, under the head of "*Female Diseases.*"

We hardly need remind the reader to bathe, wash and attend the child altogether according to the strictest rules of Hygiene; the use of a cap, for instance, on the head, by day or night, is considered entirely superfluous, as it proves oftener detrimental, than beneficial. Equally pernicious to the physical welfare of the child is the too tight bandaging of its breast, limbs or abdomen. Every part of the body must be left as free as possible, to enjoy full liberty of motion, an especial condition and means of health. The child must exercise its limbs and lungs to the fullest extent. Nature wants it; science demonstrates it, and unconscious instinct, as exhibited, for instance, in a young infant, establishes exercise as a law of nature. The child has an instinct, which governs its appetites and desires, frequently more correctly than it can be expressed by language. Let us only carefully observe its wishes and not stifle them by putting the child's body in a straight-jacket, wherein it cannot express what it wants. Nature herself, takes care of the child, we must only observe, not impede her dictates. Hear the beautiful and graphic description, by Dr. Eliz. Blackwell, in regard to the care which nature constantly extends to her little darlings, for protection and healthful growth.

"The young infant is almost withdrawn from our control. Nature says to us, 'stand by, and watch my work!' This delicate life will admit of no trifling, no neglect, no experiment; but watch the infant, how it kicks and cries, and works, not arms and legs alone, but every part of its body in pain or pleasure. We sit and smile or silently weep; but the baby puts every muscle in motion; if it is pained or angry, it will scream with its whole life, and contract every little fibre, and strain and wriggle in infantile rage, to the intense alarm of its mother. We may leave it to nature for exercise; it will be well attended to, and carried through an efficient course, reaching **every muscle of the body,** that we should find difficult to imitate by art. Watch the little child, too, that has learned to walk and prattle. Do we need a more perfect illustration of perpetual motion during its waking hours? Give it free room and a few playthings, if they are only blocks of wood, and it will go through a series of positions, stooping, twisting, doubling, turning over, that are incalculable and unapproachable. And you cannot quiet such a child; take away the playthings, and every legitimate source of amusement, and your inkstand will be upset, your books ingeniously torn, the table-cloth dragged off, and the contents of the work-basket sent rolling; and if it be absolutely restrained from such questionable devices, it will make it up by fretting and fidgeting till the older head fairly aches. It is a most admirable arrangement, this incessant activity of the child, the inexorable law by which it lives, and which will turn the whole household upside down, sooner than sin against its own nature. For it lives by movement; fresh air and exercise are the mainsprings of its healthy physical life. Thus in the *earliest* years of life, nature's indications are very plain; and in exercise, as in the organic functions, the most perfect freedom, under favorable conditions, should be enjoyed by the child, that its own instinct may guide it."

In a good foundation, lies the main strength of a superstructure; even so is it with a good constitution, the basis of which must be laid in earliest infancy. And yet how often is it neglected by parents or those in charge of children. Science has furnished the farmer and husbandman with strict rules in regard to the rearing of good stock; these regulations are followed to the letter, and enterprise is thus an hundred fold rewarded for its expense and trouble. Science, also, has prescribed the best method of attending to the physical education of children, but how few follow its dictates. Those who have done so, have reaped the reward in rearing strong and healthy children, at once their delight and a blessing to their country. But too many, as yet, who behold such praiseworthy examples, consider them exceptions, freaks of nature, or the consequence of accidental good fortune. This is a sad picture, but a true one. We must have indeed strayed far away from nature's own path, to consider a pair of rosy cheeks or a lively, energetic disposition in a child, nothing else but a freak of nature. So scarce has that become which ought to be universal!

The fault most productive of those evils, has been the great physical restraint under which we put the child as early as the first day of its existence. Tight bandages, compressing the abdomen and breast, are applied immediately after its birth. Afterwards it is closely confined in heated rooms, not allowing a sufficiency of fresh air, so essential to the development of the young organism; its brain is heated by a cap, its stomach deranged by improper food, which is forced upon the little sufferer, and if pain and restlessness follow, paregoric, Godfrey's cordial, etc., must restore quiet, or castor oil remove the evil. Still greater distress follows such violent and senseless treatment. Congestions to the head appear and convulsions threaten. No wonder that the constitution of a child, under such mismanagement, can never be a strong one, even if it survives the attacks that occur during early infancy. But how many do survive? Only one half of all children born, reach the age of two years. This is a melancholy fact, casting a dark shadow, freighted with destruction and death, upon our so-called modern civilization, with its boasted light and instruction. If one-half of the human race has to perish before reaching the second year of its existence, we have not yet begun to realize the blessings of reform in our treatment of the helpless young. It is full time that we should put into operation the measures devised by science, and calculated to keep the angel of death from the cradles of our children. These are at once comprehensive and effective.

If we see our faults, let us forthwith correct them; no time is to be lost. We have perpetrated sins of commission and omission; we must know how we have done wrong in order to understand how to do right. We have wronged the child in omitting to give it a sufficiency of fresh air, water and exercise, all of which are indispensable conditions of its thriving well and receiving a firm, healthy constitution. But more than that, we have wronged the children by actually making them sick, committing an attack on their life and constitution, by compressing the lungs and other noble organs, rendering them, thereby, weak for life. We also prevent the liberty of motion by bandaging the limbs; we irritate the intestines by castor oil, and debilitate the brain and nervous system by opiates; and, finally, we over-stimulate mental action by too early application in that direction, merely to gratify our pride and foolish aspirations; a wickedness often enough punished by the early death of the object of our love and hope. Let us avoid these faults, and not one-half of the dear, helpless beings, will fall victims in early life, while the surviving majority will be blessed with a healthy constitution.

Hygiene has become a science and demands a treatise of its own, so extensive is its range of action and practical utility. We have not the room or intention to give here its details; the reader will find these in separate works on Hygiene. We content ourselves in pointing out the grosser faults at present committed in the rearing of children, and their remedies. And as the female infants suffer, in proportion, more than the male, we consider the above remarks especially justified. The female sex ought to have, above all, healthy constitutions, being destined to play the most important part in the propagation of the race. Let us raise strong and healthy mothers, and there will be at once an improvement in the health of the next generation. The female, therefore, requires at our hands, the most careful attention, and we are bound to commence it at the earliest period in infancy.

When vaccination shall take place, let the female infant be vaccinated, either high up on the arm or on the outside of the leg, below the knee, (always the best place for small infants,) in order to avoid the scar from being observed afterwards, as the girl or young woman frequently appears with bare arms. It can easily be done, thus preventing an ugly scar from marring the beauty of a well formed, symmetrical arm, no small attribute of the physical perfection of woman.

Before we proceed farther, to discuss the best methods of education for the young girl, we would express at once our decided condemnation of those at present in vogue. We have for a long time witnessed the bad effects which the educational system now adopted in most of our Boarding Schools, has on the health and minds of the daughters of the land. It is entirely erroneous, and mischievous in the extreme; wrong from the commencement, its results cannot be beneficial.

The object of nature is, to prepare the system in the preceding period for the next one following, in childhood for youth, in youth for womanhood, etc. We must, in our educational efforts, observe the intentions of nature, and not pervert or overleap this order. To teach a child what a youth ought to know, and so on, or to neglect or prevent the development needed for a child, in order to make it perform the duties of a youth, will be an injury which never can be fully repaired, as the more advanced period can never acquire that which should have been the object of its pursuit in a previous one. Each period of life has separate uses, which must be fulfilled, and which never can be changed without serious derangement; this is at least the general rule, the order of nature.

The development of the *physical* system occurs principally during childhood, which extends to the age of twelve or fourteen years. During this time our endeavors should be directed almost exclusively towards the support of the physical growth; we must at least refrain from interfering with it.

All education is properly divided into two parts, analogous to the two-fold existence of man, physical and intellectual. The moral preceding and following the latter, is therefore included in it. Each of these two departments has separate ends to accomplish, and will be required in different periods of life. The physical training has for its object the education and strengthening of the body, in the whole, as well as in all its parts. The body is the carrier and instrument of the mind; its strength and health are all-important for the easy and complete performance of the real or spiritual life. To make physical education effectual, we have to commence it in early youth, and pursue it steadily during the whole period of bodily development. This period, in fact, ought to be filled up almost exclusively with the practices necessary to carry out the principles of a thorough physical education, else the succeeding stages of life will result in fewer advantages for the object in view. We are firmly persuaded that the greatest blessings would flow from following the above principles in our common school system. Their adoption would not interfere with present arrangements, as we need only to alter the objects and hours of instruction. The child may, as heretofore, be put under school training at the age of six, but from that up to eight, it should receive twice, every day, half an hour's instruction in the purely elementary branches of education; the other time should be devoted to a regular and systematic practice of gymnastics, under the superintendence of competent teachers. From eight to ten years, the next higher branches should be taught during two hours in the day, while more difficult gymnastics should be pursued during the remaining time. From ten to fourteen years, a confinement of four hours a day in the school-room, for still higher studies, would not be prejudicial to the child's health, there still being left four hours a day for active gymnastic exercise.

Such ought to be the instruction of the young, in order to make the body strong, while its expansion takes place, and to train by degrees the mind to those exercises which afterwards will be the principal business of its life. We are wrong, if we suppose that the great object of school instruction consists in filling the head of the young with actual knowledge. The acquisition of this ought to be of minor importance; sometimes it is really detrimental. The school can only *train* the mind in the paths leading to knowledge and thought, to enable it to improve by its own exertion and observation. This is an important truth, proved by the experience of great men, who were for their greatness indebted not to the actual knowledge taught in schools, but to the impulse which their minds received by the training of those schools. We therefore, strongly recommend parents and teachers, not to subject the young and elastic system of the child to the cruel confinement of the school-room, but to educate the body, rather than to fill the mind with premature knowledge. A child of ten years of age can learn in a quarter of the time what is offered to one of seven or eight; it will be an easy task for the older, while the younger child will suffer and labor hard, to accomplish the same. There is no time, therefore, gained by hastening with the intellectual education of a child; let its brain first mature and harden, easily to perform intellectual labor. Besides, it is not necessary to accelerate the intellectual development, in our days of rail-roads and ready intercourse, which offer to the young an easy medium of instruction not heretofore known. A child can learn to read and write, by merely having its curiosity and imitative faculty excited, through the innumerable hand-bills posted up in streets and thoroughfares, on steamboats, and on rail-roads. The means and objects of observation being increased an hundred fold, compared with former times, there remains less for the school to instruct.

The physical cultivation is, therefore, the first we have to look to in a child. It comprises different practices, from the mere running about in the street, to the most complicated gymnastic exercises. It forms a complete system, and ought to be taught and practised in perfect earnestness. We will dwell upon the more important exercises at some length.

The girl should be permitted to run about in the open air, and exercise at least as much, if not in the same manner, as the boy. Her plays are naturally different, but the benefit derived from them for the development of the system, is the same. If the boy flies the kite, the girl rolls the hoop and jumps the rope. The latter amusement ought to undergo a surveillance by older persons, to avoid an excess of action and consequent injury to the nervous system, which is naturally excitable in the young. The girl should be exposed, even to the same degree of inclemency of weather as the boy, in all seasons, in order to harden her system and prevent precocious development. This latter consideration is even more weighty in her case, than in that of a boy, who develops more slowly, while she is inclined to precocious development. To counteract this more fully, frequent cold bathing is necessary, which, having become a habit with the child, will continue to be enjoyed as a luxury by the girl and maiden.

The dress worn by the girl at all times, particularly during play hours, ought to be made to fit loosely. Any pressure in this age has a very injurious effect on the physical development, impeding easy, free and abundant motion; compressing the bony structure, and thereby preventing the nobler internal organs from expanding. Spinal and lung diseases are the frequent result of a violation of this rule.

Another very excellent exercise, suitable and natural for the girl, is dancing, the artistic rules of which, as an accomplishment, may be acquired in this period of life, better and sooner than afterwards. The child between seven and fourteen years, is naturally inclined to exercise, will be, therefore, fond of the dancing school. Its movements can be directed, in this age, very easily, because the mind of the young is less fettered by conventionalities and restraints, which produce so much awkwardness in after life, if one is not regularly instructed in the free and graceful motions of the body.

In thus recommending parents to have their children, and particularly their daughters, acquire this beautiful accomplishment, we do not advocate its excessive practice among the young or adult. We have seen moral and physical evils resulting from its abuse, and feel very anxious to warn parents, not to allow their daughters to become too much fascinated by the pleasures of the dance. The best of every thing may be abused; this should not prevent us from using it moderately and to the purpose. Dancing, systematically taught and rationally pursued, improves the beauty of the natural gait, and is conducive to bodily health and a fine flow of hilarity and enjoyment. We protest, however, against all fashionable nonsense, by which dancing, as an art, is carried too far, and becomes too difficult for easy instruction or quick acquisition. It ought to be taught in a simple, easy manner, without the affectation of the modern fashionable dancing school, in which this liberal accomplishment is presented to the pupils in a professional perfection, not suited to educational purposes. The latter object is only needed and desired, which to effect, nothing more is necessary than to teach the child the various graceful attitudes and motions, having reference to the improvements of its own natural walk and carriage.

But it is not necessary only to establish by art, the elegance and grace of the human figure; we must also endeavor to raise its physical strength. For that purpose, the systematical exercise of the muscles, as taught by gymnastics, is strongly recommended. The term "gymnastics," signifies physical exercises, according to scientific rules. Now, many might believe it needless to take exercise under the control of certain rules, thinking that bodily exercise in any shape or form, was the same in its results, viz: strengthening the system. This, however, is not so. Exercise may do harm as well as good, and great discrimination is needed to apply its force at the right time and in the proper direction. To let a person with feeble lungs take exercise in walking every day for a long distance, would certainly not improve his condition, it would rather **be injurious to** him; while exercising the muscles of the chest by throwing out and drawing back his arms, would be beneficial. Thus every muscle in the whole system has its own beneficial effect, and should be strengthened by an appropriate exercise. For this purpose to put successively all muscles in active motion, the art of gymnastics has arisen, which teaches the various methods of doing it.

A double benefit is gained by this process. It is not merely the acquisition of physical strength resulting from these practices, but also the training of the mental faculties, indirectly coming into play during these physical exercises, when performed scientifically. The pupil is constantly reminded that the use of certain means will be needed to accomplish certain ends. Thus, his faculties of calculation, decision, energy, order, etc., will be intelligently acted upon. The unscientific exercise of the body is as different from the scientific, as the playing of one who is unlearned, on the piano, compared with that of an experienced musician. The one produces an unharmonious, offensive noise, while the other delights and instructs by his performance. An important art, of so vast a range, cannot be taught and practised sufficiently in a short time. To be accomplished in the art of playing on the piano, requires years of patient, persevering effort; and the body is analogous to and even more complex than a piano, having four hundred muscles to be set in motion or played upon. It must require years of continued exercise and study to bring these hundreds of muscles under intelligent control, to act upon and with them, and thereby improve their several conditions and strengthen the whole system. The introduction of gymnastics ought to be commenced, therefore, in early childhood, while yet the body easily yields to, and even delights in exercises of all kinds; their study can be made very attractive to the youthful mind, as these practices not merely delight the bodily senses, but also engage and invigorate the intellect, acting constantly upon the child's attention and discrimination, that it may comprehend the different motions, and distinguish one from another. A child cannot well be put under gymnastic training before it is six or seven years of age; it will then be sufficiently advanced to understand and retain the lessons, which begin like other systematic studies, with elementary efforts, and advance gradually. With the growth of the pupil, the exercises increase in power and variety. Every voluntary muscle of the body is, one by one, acted upon, and brought under the control of the will; particularly is this the case, where parts of the body or systems of muscles seem to be weaker or less developed than others. Regulated exercise increases their strength and bulk, and restores thereby that harmony of organic development, so essential to the future health and permanent welfare of the whole system. Thus the gymnastics assume an hygienic importance, not equalled by any other means, to counteract or prevent disease. And this in a greater degree for the girl than the boy; because the former, naturally weaker in frame and muscle, suffers more from early confinement in the school, and becomes thereby particularly inclined to spinal and lung complaints, these scourges and tortures of the female sex. Having arrived at this part of our subject, viz: the hygienic bearing of gymnastics on the present and future welfare of the female system, we would call the reader's attention again and again, to its vast importance, by reviewing the present treatment of girls during their education, and its bearing upon the health and welfare of the child.

Look at the lively little girl, running about all day, fixing dolls or playing otherwise in the house or out of doors, talking incessantly, and putting herself into all imaginable shapes and forms, expressing thereby her inner feelings or wishes, and exercising her muscles instinctively. Scarcely four years old, you confine this lively and lovely little creature into an infant school,* where almost the

---

* Some writer in " Chambers' Information for the People," goes still farther in torturing the poor little ones; his advice, if acted out, would certainly kill the whole infant world. Hear his

whole day long she has to sit quiet or be restrained at least in her movements and plays. Her brain, as yet very soft and impressible, is excited too much by the studies in the books, while at the same time, the general system is thwarted in its development by confinement and rest. This is just perverting the order. In this age the brain needs rest, at least, not more action than the natural instinct of the child will demand, while the muscles and bones need all the motion they can get, to develope strongly and perfectly. It is impossible to reverse the natural order of things, without suffering the penalty following such an offence. Nature and medicine will cease to cure, where the offence was too great, disturbing the fundamental arrangements of the system. The brain is proportionally larger in infants than in adults; the head therefore, does not grow as much

---

advice: "*From six to fourteen years of age.*—In a rightly arranged and complete course of elementary, intellectual education, it is presumed that the period from *two* to *six* years of age, has been spent *in an infant school.* The effect which such a preparation has in facilitating the subsequent operations of the teacher, is so great that *every effort* should be made, to give children the advantage of it."

The practical impossibility of being carried out, saves the above advice from any comment on our part; it is too extravagant bordering on the ridiculous and insane in its demands, and therefore harmless; yet the reader will perceive the danger of their infants being killed by methods like the above, recommended, or already in vogue.

or as fast as the other parts of the body. The brain being as yet very tender, easily yields to the pressure of the blood in its vessels, which predisposes to dropsy of the brain, acute or chronic, both equally fearful and fatal. Precocious mental development will over-excite and congest the brain and thereby cause inflammation and dropsy of the brain, convulsions, weakness of muscles and bones, rachitis, etc.

The young system wants air and exercise. Without these, its growth is stinted at once; it cannot bear to be shut up in rooms or confined upon benches; it needs the fullest liberty. The injury done to those innocent little ones, by sending them to infant schools, is immense and can hardly ever be repaired, as the injury is inflicted during this early period, mostly on the spine and breast bone, often also, on the pelvis, causing rachitical diseases, which positively destroy the best part of life's happiness and destiny. Parents, this misery could have been spared to you and your daughters by a little reflection and action, in the right direction with nature, not against her. It is true, your excuse is perfect; you did not know any better. You did as others have done, and still are doing; you could not be blamed, neither were you aware of any harm being done to your darling. Did the little daughter not come home from school cheerful and delighted with her school-mistress; she loved her so much, she could not now stay away from school, she would be very unhappy, etc. You are pleased to watch and see the great progress your child makes in reading, writing and cyphering; what a capital teacher she has, and how forward the child is; there never was such a child. Its future is speculated upon with no little relish. Poor parents, all this time you have been striving with all your might to ruin the health and prospects of your child, nay, even the intellectual developments which you intended to foster so early, and thoroughly, you have impeded, thwarted and stinted forever. But, you say, why is it that the child, if not naturally inclined to intellectual pursuit, loves it so much that it sometimes even cannot be persuaded to leave the books for playthings? This is very obvious, if we consider the means which are used to make the child fond of books and study. Not to mention the fact, that children like to be in the school-room, because they prefer the society of their own age to that of a more advanced one, its noise and excitement to the quiet and restraint of home and its parlor; there is another still greater inducement to draw them to books, instead of plays, and this is one of our own fabrication. The ambition of the young mind is stirred up in that early period of life, in order to arouse its energies to study, and hard intellectual labor.

Thus ambition, this legitimate stimulus of a more advanced age which needs its sting and propelling power, is used by our enlightened teachers, to set the brains of children on fire, and put their intellects into hot-houses, before their hearts are prepared to expel envy and malice, almost always the inseparable companions of ambition. It is dangerous to arouse passions, even the noblest, given to us by Providence for wise purposes, before the time of their natural appearance has arrived. Be these passions either of a moral or physical nature, their harmonious workings have to follow the same laws. If aroused prematurely, their action becomes destructive by engendering morbid conditions and precocious development. But if allowed to lie dormant in the system until the time appointed by nature, they spring up in healthful action and vigor, accompanied always by their counterpoises, antidoting and restraining their activity if too abundant. In this manner, if ambition should become too great and unscrupulous in a full grown man, prudence will arise to curb its impetus; man will reflect on the evil consequences which may result from a too ambitious desire and thus the equilibrium of his mind is at once restored. Not so in a child, where reflection has not yet appeared to restrain the will if under ambitious influence. Children, whose ambition has been unduly stimulated, sometimes have received serious injury in body and mind from such unnatural races; they frequently have died from diseases, thus contracted. A medical friend at my side just now relates to me the sad story of his sister, who lost five of her children, all during the first school period, from precocious intellectual development; the sixth one was saved by adhering to the Doctor's strict rule, not to foster mental but physical development.*

It was the fault of former times to educate the young mind too little; we have fallen into the other extreme of educating too much, by over-taxing the minds of the young. The middle course must be kept, otherwise the harmony in

---

* I am perfectly convinced that harm is done by the premiums and prizes offered at the examinations in schools, to the most forward pupils. It is a system which naturally had to accompany the hot-house education, as it exists at present everywhere, and for which it furnishes the best fuel to force the tender plants into premature mental growth, regardless of sound physical basis. If modern educationists had offered to the mind of the child *attractive* and *comprehensible* studies, they would find the stimulus of gain or preferment, in the shape of prizes, premiums and places, unnecessary to induce the child to exercise its mental faculties. Moreover, it is wrong to make the child labor, sometimes above its natural powers, by holding out these inducements to his young, lively soul, thereby inflaming the lower passions of gain, pride and ambition, and poisoning its harmless, innocent existence with the bitterness and stimulus of an older age; vices of the same age will find an open door, and thus it is, that at present we frequently find our youth having become old, before they have been young.

the double nature of man is disturbed and his growth, in either direction, stinted.

We have above alluded to only one of the many erroneous practices into which modern educationists have fallen by attempting to raise the standard of education. We have shown that it is not in harmony with a child's development, to arouse in him, prematurely, one of the most powerful passions, without being able to bring into play its counterpoise, and that it is dangerous, even criminal, to do so, as frequently, thereby, diseases are provoked fatal to life and health. We now will add, that it is also cruel to do so, because the ambitious child, having strained its mental powers to the utmost, and still not being able to compete successfully, meets thereby with one of the most agonizing draw-backs which falls to the lot of man. Have you never seen the burning tears rolling profusely down the child's cheek, heated with shame or rage, after an intellectual race was lost? These juvenile disappointments, the frequent and natural results of our present educational system, are as keenly felt, and as bitter in their taste, and as hardening in their after-effect, as those experienced in later years, and perhaps more so, as their severity is not softened by reflection or prospective reparation. The bitterness of the moment is felt in all its disagreeable power, and often crushes, at one blow, the aspirations of the young, their hopes and energies. You may say that all this only lasts for a short time, that young blood soon forgets the ills of life; yes, it may be so, but nevertheless the young heart is deeply wounded, and although the wound soon heals, a scar will be left to impede the natural and wide expansion of the heart. We ought to be very careful not to offend, without good and wise reasons, the child's mind, because its sensibilities are finer and more perceptive, while reason and reflection are not yet very strong.

We hope to see the day when a closer analysis of the peculiar organization of the juvenile mind shall guide those to whose care its education is intrusted. We are sure that after mature consideration of the subject, they will find it necessary to adapt their system to the developing, not to the developed child; that they will treat the child as such, and not as if it had all the fully developed faculties of the adult.

A general school system should be adopted, based upon the above principles, viz: the physical education by means of gymnastics, dancing, music, etc., together with instruction in the most elementary branches of knowledge up to the tenth year of age, and afterwards, up to the fourteenth year, the higher branches of instruction together with higher gymnastics. Were such a system adopted, we should soon enjoy its good results; the next generation would have strong, intelligent mothers, capable of filling, in the full sense of that word, the responsibilities belonging to them. It is the sacred duty of every one comprehending the vast importance of this subject, to work for its realization.

Above, we have mentioned music as one of the elementary branches to be taught to children. It is necessarily comprised in a catalogue of instruction. Music, in its composition and effect, is emphatically the most humanitary of arts. It belongs to all countries, races, sexes and ages; it enters into every one's organization; its harmony and melody are the very soul of all that exists, and its tact and rythm form the mathematics, the crystallizing principle of the world. We cannot have too high an appreciation of its value, either as a source of enjoyment, or in the culture of body or mind. The child perceives and feels its influence as readily as the adult reads, in its swelling notes, the highest thoughts and sublimest sentiments. Music is universal in application and effect, the best introduction, therefore, to all other sciences and branches of knowledge. To exclude music from the schools would be a death-blow to all education; its fertilizing principle would be wanting, leaving a waste in the soul as barren as the sands of Sahara. Music, like morals, must be taught, practiced and enjoyed during our whole life, commencing with the earliest dawn of perception.

Girls, above all, should be well instructed in

music; not to make of them professional singers or players, but to let them enjoy and study harmony and rythm, the constructors of happiness and bliss. The young soul and hand of a girl is ready to cultivate music, particularly if her mind be not over-burdened with premature knowledge of a more abstract nature. We recommend, therefore, instruction in music during the whole time devoted to education. For young girls, the piano forte will be the best instrument for instruction; in after years the harp becomes an elegant and appropriate means of further musical study.

The voice ought to be cultivated at the same time. Singing being a natural gift of the female sex, it would be hardly necessary here to admonish parents not to neglect its proper cultivation in the education of their young daughters.

Having, thus far, considered the physical and intellectual condition of young girls, it remains for us to notice their *moral* and *religious* training. But as this comprises the most important elements of individual happiness, and as such must be left to the especial care of the child's natural guardians, we refrain from mentioning the means, necessary for the accomplishment of this object; they are known to, and within reach of every one, as we live in a Christian country.

One remark may not be out of place here. Religion, like music, is universal, and the very soul of our being; let this soul once awake, and a new creation will appear, immeasurably exalted above all others. And as religion, like music, can be enjoyed doubly ir sentiment and thought, it follows that its teachings can and should be commenced, like those of music, early in childhood, when the finer sensibilities of our nature are yet in full play, and religious sentiments fasten the attention of the young soul, as the soft tones of the Æols harp.

Before we conclude this chapter on woman's girlhood, we will say yet a few words concerning the means of education at present in vogue. We have already mentioned that the popular school system is altogether defective as regards rational education. It might be thought, however, that female Seminaries and boarding-schools had obviated these evils and replaced them by better educational means; but this is not so. They are even worse, in many respects, than the common schools of the country. Their plan of education is almost wholly based upon the most rigid intellectual training, destroying the physical system in the very bud. The number and variety of studies imposed upon a young girl in these institutions, is really frightful if not ridiculous, each establishment trying to out-do the other in these particulars. The programme must be full, comprehensive and novel; if so, it will draw pupils. Hear Eliz.

Blackwell, M. D., who, on this subject, must be set down as good authority:

"The most abstruse subjects, that tax the attention of the strongest mental powers, are presented as studies for the young; girls of thirteen or fifteen are called upon to ponder the problems of *mental and moral philosophy*, to demonstrate the *propositions of Euclid*, to understand the refinements of *rhetoric* and *logic*—admirable studies, truly, but they are the food of mature minds, not suitable to children. But it would puzzle the most ingenious observer, to discover the *good use* of most of our children's studies. If the object be mental discipline, there is no surer way of defeating such an object, than to attempt to give the mind a superficial view of a subject too difficult for it to grasp—to confuse it with a multitude of disconnected studies—to hurry it from subject to subject, so that the simple studies more suited to the young mind, are imperfectly acquired and soon forgotten. * * * How can it be otherwise, when the young mind has to apply itself, during the limited term of school study, to such a list of subjects as the following: Grammar, Ancient and Modern History, Natural Philosophy, Chemistry, Botany, Astronomy, Mental and Moral Philosophy, Physiology, Rhetoric, Composition, Elocution, Logic, Algebra, Geometry, Belles-Lettres! Now for the accomplishments: French, Latin, Italian, perhaps Spanish, German and Greek—I believe Hebrew is not introduced in this country—vocal and instrumental music, piano, harp, guitar, drawing, painting, and various kinds of fancy work."

This is truly a formidable array of studies for a young girl, while she is at boarding-school. It is impossible for her to do full justice to all of them; the attempt would certainly prove fatal to health. But, thanks to the native sprightliness of youth, they slight most, if not all of them, and thus manage to escape with their lives from the ordeal of the fashionable boarding-school. It is true they have acquired a smattering of knowledge and perhaps outward polish of manners and accomplishments, but it is only superficial, imperfectly acquired and soon forgotten. Has anything been learned, really useful in after-life? No, absolutely nothing, save perhaps reading and writing. And, as regards the first, her taste acquired in the boarding-school, may be anything but the best. Has she secured a healthy, strong body, to sustain her in the duties and cares which will soon follow? No, her body is perhaps less strong and healthy than when she entered the school. How useless, then, nay, how ruinous, must be a system of education, which promises so much and effects so little good, aside from the real evil it does. The world never was punished with a worse educational system, or one which so completely annihilates the hope of the parent and philanthropist, by nipping the welfare of future generations in the bud. It is full time for this nation to look well to its educational matters; because their influence is vast, and the most important interests are at stake. Reform in this respect, should take place soon, or the future flower of the nation will be withered ere yet it opens. With a total change in the course of studies, and the adoption of gymnastic instruction, the young girl will receive that kind of education which will fit her for the severer duties of after-life.

CHAPTER. III.

MAIDEN, OR YOUNG LADY.

Just before or about the time, when the girl becomes a maiden, or as we now say, a young lady, great alterations have taken place in her physical system; changes, the nature of which will, from this time onward, affect her whole future for evil or good. At the same time the girl's mind receives a new direction; she behaves differently, is more reserved in her conduct, and more careful in her appearance before others. The *monthly period*, or *menstruation* has appeared.

As it is important for mothers to know the use and signification of this periodical discharge, which is the herald of such vast changes in the female economy, and the balance-wheel of her health during the most eventful period of her life, we will treat of it in these pages more fully, showing its origin and connection with other vital processes, occurring simultaneously in the system As in the course of explanations for this purpose, it will be necessary to make the reader acquainted with the most secret and sacred proceedings of nature, we will endeavor to convey this information in a manner as little exceptionable as possible.

A mother should have a correct knowledge of these processes, so intimately connected with her own and the happiness of her daughters, dispensing continually either health or disease. If it is not given to her correctly, her natural curiosity, impelled by the interest attached to the strange phenomenon, will incite her to procure from other sources, wherever she can, such knowledge on this subject as may, perhaps, lead her into misery and danger.

Before the true nature of menstruation, its cause and object was fully known, the treatment of female diseases, depending on menstrual disorders, was very uncertain and hazardous. Recent investigations have dispelled this doubt and uncertainty; the anatomist and physiologist have combined to reveal to us these most secret transactions of nature, thereby enabling the physician to prevent and cure their diseases more successfully. Information so important and useful, should not be withheld from the people, if we can correct thereby the many erroneous opinions which are yet afloat respecting menstruation, its origin and signification in female development.

The menses have been generally considered the surest evidence of a girl having passed into the state of womanhood. In most cases, particularly in healthy girls, this is true. But as menstruation is not properly the cause of this change, but only an outward sign of those internal preparations necessary for such a change, it may frequently occur that, by some morbid conditions, this periodical discharge does not make its appearance, although, from other unmistakable signs, the girl has become a young woman.

These signs which, besides menstruation, indicate the approach of womanhood, arise commonly between the twelfth and fourteenth year. Some are invariably present, such as the increased size of the hips and breasts, the roundness and swelling of the limbs, the perceptible fullness of the whole form; others are variable in their occurrence, having reference more to mental development.

The girl, talkative, roguish and romping, becomes at once reserved, retiring, sometimes even sad and easily moved to tears. She begins to dress with more care, and is more observing anxiously and silently; her whole soul is filled with gentle emotions. She longs to enjoy the pleasures of sincere, disinterested friendship, that love in a bud, which makes life a delightful journey. In this state of mind, she clings to her mother for advice and counsel, showing more affection towards her than heretofore. Then it is that a mother can exert the most beneficent influence over this developing, interesting creature, just ready to become a woman. In this time of fear and hope, the mother ought to instruct and counsel her as to the meaning and import of the various phenomena, so strange to the trembling girl, but of the utmost importance to the developed woman. At first, naturally modest, the girl will often hide from the sight of the mother, what she considers to be singular and wrong. The first appearance of the menses may even frighten the timid girl, who does not know the meaning of such an occurrence. A watchful mother or female friend, should never fail to give the girl such instruction and advice as will dispel her fears and guide her actions.

A general development of form and size takes place mostly from four to six months, prior to the first appearance of the menses. This first outbreak, like the teething of infants, is frequently accompanied by many morbid symptoms, principally of a nervous and congestive nature, such as head-ache, palpitation of the heart, sensation of smothering, irritable, quick and impatient temperament, sometimes followed of a sudden by sadness and depression of spirits, restless nights, pains in the small of the back and loins, etc. These symptoms disappear as soon as the discharge commences, which may last at the first time for two or three days. The proper average duration of the courses afterwards, is five days; if it is below or above this standard, it is caused by morbid conditions or other modifying circumstances.

As it is all-important for the health of the girl to have the menses established well and regularly, it is necessary not to overlook, in the beginning, those morbid symptoms, above alluded to. They may be ignored once or twice without producing serious injury; but if the menses continue to be accompanied with pains in the back, cramps in the stomach, etc., we must not neglect to call in medical advice. We refer the reader to Part II., where, under the head of "*Difficult Menstruation,*" the remedial course to be pursued in such a case, is indicated. The nature of these pains and apparent obstructions will be understood after a consideration of the internal proceedings connected with menstruation and its object.

Puberty in a female, or the aptitude of becoming a mother, is produced by the action of two small bodies, lying on each side of the uterus, and connected with it by small tubes, leading into its cavity. These small, oval-shaped bodies are called *ovaries*, or *egg-beds*. They are composed of a formative material, called *stroma*, which contains small vesicles, ova or eggs, the construction of which, in all its essential parts, is similar to the common egg, when yet without the external shell, even as regards the presence of the yelk, the main dependence of the growing germ.

Their size, however, is extremely small, not exceeding that of a pin's head. The same wonderful arrangement obtains in the vegetable kingdom, which also propagates its kind by means of ova, or little eggs, as far as is known at present. For these astonishing and interesting discoveries, we are mainly indebted to the microscope, by the aid of which the exact structure of the various parts engaged in these proceedings was fully revealed, and their function determined. We will presently see with what precision and harmony the different organs co-operate to facilitate the propagation of the species.

The little egg, which we will call hereafter ovule, lies dormant, enclosed in a sack, until it becomes stimulated by the reproductive power of the system, which, as we now know, returns periodically in the human female about every month. At that time it begins to grow, bursts the sack, and escapes into the Fallopian tube, which carries it into the cavity of the uterus. In single life it passes away with the menstrual discharge, which, as a secretion of the uterus, is just then excited by the same reproductive power of the system. In married life, when the conditions of nature are fulfilled, the ovule is retained within the cavity of the womb and there developes to a perfect human being. The explanation of this process will be given in another place of the book. For us it is, in this connection, important to know that menstruation stands in the closest relation to the generative power and process of nature; that its healthful appearance is indicative of a perfect development of these forces, but that a premature hastening, or a tardy appearance of the menses by disease or artificial means, must injure the above named functions and thereby the whole female system, formed, in a physical respect, especially for that purpose.

As already stated, menstruation appears about the fourteenth year; this may be considered the normal standard, although frequent deviations from this rule may take place. If it occurs earlier, diverse circumstances may have hastened its appearance, such as luxurious habits, indolence, sensual indulgences, reading of novels, etc.; also sedentary habits and too close application to study, have a great tendency to produce menstruation in advance of the other signs of womanhood, mentioned above, which must precede the menses.

A girl must have become, first, broader across the hips, the breasts must have enlarged, and her form filled up in rounded outlines, before a healthful menstrual discharge can be expected. If this is not the case, the cause of the non-appearance of the period lies in the backward ovarian development, which generally has its foundations in the qualitative deterioration of the blood. This then has to be ameliorated before the menses can appear. The remedies to effect this will be indicated in the Part II.

From the above it will appear how useful, in a practical point of view, these discoveries have become, directing our remedial means to the places really diseased. A physician, without a clear comprehension of these secret occurrences would be unable to treat successfully their disorders. Action without knowledge becomes frequently fatal. An instance of this is related by Dr. Dixon, as follows:

"We have seen cases, in which mothers demanded importunately medical treatment for children, possessing not a single sign of womanhood; and upon one occasion, in which we very unwisely refused to prescribe for a young girl, death was the consequence of a powerful medicine administered by a well-meaning, though ignorant parent. In this instance we might, by apparently yielding to the parent's desire, or by prescribing some harmless drug, have gained time, as recommended by some humane physicians, until menstruation was produced by the effort of nature. The case made at the time a strong impression upon our sympathies, and we determined to use our humble powers of popular instruction, when time and experience had given us more knowledge of the subject."

If menstruation is once established, it generally

returns every twenty-eight days. Its duration each time is about from five to six days. Its too early or too late appearance depends mostly upon morbid constitutionality, engendered by heritage or wrong habits, by diseases having a special reference to the uterine region, luxurious living, etc The amount of menstrual discharge varies very much. The normal standard, however, may be set down safely at from five to eight ounces during each monthly period. Yet, this can not rule individual cases, which are governed by peculiarities in constitution and habits. If the woman remains healthy, the object of nature is fulfilled.

The same constitutionality governs the time of appearance, and may modify the normal standard of twenty-eight days frequently, without inflicting any injury upon the general health.

Having thus dwelt at length on the nature of those proceedings which, silently preparing for action during childhood, break forth at last in the monthly period, as the surest outward sign of the important change from childhood to womanhood; having considered its origin, use and effect in the female system, we are now prepared to follow the young woman in the different spheres of her activity. She has not alone changed her physical appearance, as we have seen, but her moral nature also differs essentially from that of the child.

While a child, that is before the appearance of her monthly period, her mental faculties were less engaged with the proper, nice and decorous. She was unobserving, careless as to drawing the attention of others towards her; she yet participated to a certain extent, in the free, romping, even wild character of the boy. She had yet to be governed by others, parents and teachers; the rule within herself, that priceless jewel and powerful weapon of a female, womanly modesty, had not yet commenced to guide her steps and desires, which hitherto expressed themselves as mere appetites. She was still a school-girl, thoughtless, sprightly and joyous.

But scarcely has she passed the Rubicon of woman's development, when the wild, romping girl becomes thoughtful and retiring; she dresses with neatness and elegance; her gait and carriage assumes an elastic dignity; she is anxious to please and to be observed; her motions and desires are regulated by gracefulness and modesty. Although yet under the guidance of her parents and teachers, she already thinks and acts for herself; she feels that she has responsibilities and duties.

If she has been educated religiously, she will now realize more the comfort of an intelligent address to the Supreme Being, and feel the need of a reliance on Divine assistance, the more she becomes acquainted with the deceitfulness of the world. The closet will be a favored place for her, and the teachings of the Bible, which now become practical and real, will be sweet and priceless to her heart. The position, which a christian young lady occupies, is elevated beyond any other, if she feels and understands her duties in this respect.

The heart, whose throbbings she now begins to feel, for good or evil, for the glitter of fashionable society, or for the high, noble truths of science and religion—this swelling, loving heart will be regulated for the better, in its desires and loves by the mild but earnest teachings of the Gospel. From this book she should never cease to draw the spiritual nourishment of which she now stands in need, perhaps more than ever. In youth, lay up the stores for a more advanced age. The young lady should be the last to neglect the closet, with its meditations and prayers, or the public service, with its high moral teachings and devotional sublimities.

In the family circle, the maiden occupies a peculiarly interesting and useful position. She is the pride of the father, the hope of the mother, and the cherished object of brother and sister. Towards the younger members of the family she feels and acts like a mother, while they look up to her as such. She becomes, by degrees, the support and delight of her mother, the link by which the past and present are bound closer together, and are made to understand and love each other. She is now destined to become more and more acquainted with the practical duties of life. Although her studies in the school or under private instruction, are still progressing, her mind already inclines to be occupied, not alone with the abstract, but also with the real. The many little affairs of life as they occur among her associates, or in social circles generally, engage her attention and stimulate her action. Slowly, but surely, does this preparation of her girlish nature for the arduous duties of a married life, proceed, sometimes stimulated by outward excitement, (public lectures, festivities, etc.,) and inward impulses, (the dawnings of a first love,) or at other times checked and purified by the teachings at home and abroad.

At home, under the eyes of her mother, she is made acquainted with the duties of the house and family. She will be taught how to regulate the simple but various affairs of a household, which seemingly small and unimportant, confer, by being constantly and scrupulously attended to, such an attractive charm on everything connected with home and the fireside. Her own genius soon breaks forth in creating, ever and anon, new plans and devices in beautifying the home of her youth. Her fancy, order and constructiveness are constantly at work to adorn the rooms and apartments with mementos of her genial presence. She loves the home which she has thus embellished, and thereby made her own. To her it becomes the temple within which is contained all she cherishes and all she as yet desires to love.

If thus far her soul has not been poisoned by the vile fancies of bad literature, she will enjoy this happiest part of her existence, in the most perfect fulness of bliss and peace. The undefiled soul of a young maiden, presents the most lovely picture of purity and intelligence, which it is possible, in human form, to behold, provided her physical development has not been stinted by gross interference or serious accident.

The period of maidenhood is a time of physical development, and requires, therefore, a continuation of gymnastic exercises, which have a tendency to invigorate and confirm that constitution and health, thus far acquired by the bodily exertions during childhood. It will be an easy and grateful task, amounting even to a luxury and necessity, for the young girl to prosecute her exercises in the gymnasium, if her body was previously trained for them during childhood. To the readiness of execution acquired by long practice, she now adds gracefulness, elasticity and precision, which constitute the very poetry of motion.

A firm, elastic step and buoyant carriage, can

only be acquired by those whose muscles are trained and strengthened for every movement of the body. It belongs essentially to youth, and constitutes one of its most attractive charms; it is the surest evidence of health, both of body and mind, rendering its possessor most agreeable to others, and adding not a little to individual happiness and contentment. For the days will surely come, when we have to depend upon a sound physical constitution, in order to sustain ourselves under the weight of bodily labor and mental agitation. This is the reason that nature consumes so long a time in preparing, gradually, the physical system through childhood and maidenhood, until the perfect woman is formed, in all respects able to meet the arduous duties of a wife and a mother, in the full strength and maturity of body and mind.

Nothing is so fallacious and disastrous for a young lady, as the thought, that in married life, she would have no hardships to encounter and no sorrows to bear, but live in ease, harmony and quiet. If her imagination has been busy to paint the married state only as a paradise of pleasure, she will probably be desirous to hasten into it as fast as opportunities may offer; and these, in our country, are not very rare. Girls, at the present day, marry too soon and too rashly. In doing so, they not only loose the longer enjoyment of free, happy girlhood, which itself constitutes the most pleasant and joyful period of female life, but they run the fearful risk of making themselves unhappy and invalids for life, by rushing into duties and responsibilities for which they are unprepared, physically and mentally.

It is our firm conviction that the female constitution, generally speaking, is only sufficiently consolidated and established at twenty-one years of age; marriage before that time must, more or less, operate injuriously to health and comfort. But very few exceptions to this rule may be found where girls of eighteen years have acquired the physical perfection of those of twenty-one.

To prevent precocious development in this respect, mothers should guide their daughters in the selection of their social companions, with all due respect to particular predilections or favored persons. If it should be found necessary to interfere, it should be done mildly and persuasively because a mother will gain more by instruction and reasoning, than by a harsh and seemingly tyrannical course. The female mind of this country has preserved one of the finest traits which can adorn any human character, that of great independence in will and action. It is, one might say, born with them, strengthened by education, and favored by circumstances and public opinion. Against such a powerful combination of forces, the will of an imprudently harsh mother, rarely avails anything; even the wrath of a father but adds fuel to the flame. Youthful impulses, strengthened by the love of independence, soon overthrow all parental barriers, and the elopement is no sooner determined upon than it is also carried out.

But if the young girl should receive from her parents or guardians, instead of a lecture on filial duties alone, lessons in regard to her physical position, giving her a true account of the bodily development during girlhood, the inexorable laws of nature, which cannot be disregarded by any one without the most serious and even fatal consequences, we have no doubt fewer elopements or premature marriages would take place, and a great deal of misery would be prevented. The following remarks of the New York Tribune are here in point:

"The popular notions on runaway matches, fomented by the 'yellow covered' literature of the day, are exceedingly lax and mistaken. The young Miss who elopes from the parental roof to marry some adventurer who was probably unknown to her last year, is often represented as a girl of rare spirit, who does a remarkably clever and admirable thing. We hold, on the contrary, that in a great majority of cases, her elopement is unwise, giddy, ungrateful, immodest, and evinces a lascivious appetite and reckless disposition. Why should she desert and distress those who have loved, nurtured and cherished her through all her past years, to throw herself into the arms of a comparative stranger, who has done nothing for her, and whose protestations of affection have yet to undergo the first trial? It is every way unworthy of pure and gentle maidenhood to do so.

"We can imagine but one excuse for her clopement—namely, the efforts of parents or guardians to coerce her into marrying some one she does not love. To avoid such a fate, she is justified in running away; for no parent has or ever had a right to constrain a daughter to marry against her will. But where the parents are willing to wait, the daughter should also consent to wait until her choice is assented to or she attains her legal majority. Then, if she chooses to marry in opposition to her parents' wishes, let her quit their home openly, frankly, in broad daylight, and in such a manner as shall kindly but utterly preclude any pretence that her act is clandestine or ill-considered. No one should be persuaded or coerced to marry where she does not love; but to wait a year or two for the assent of those who have all her life done what they could for her welfare, no daughter should esteem a hardship.

There is some truth to be told about the 'common run' of masculine prowlers by night about garden-walls and under bed-room windows, in quest of opportunities to pour seducing flatteries into the ears of simple misses; but we have no time to tell it now. As a general rule, they are licentious, good-for-nothing adventurers, who would much rather marry a living than work for it, and who speculate on the chances of 'bringing the old folks round' after a year or two. A true man would not advise, much less urge, the woman he loved to take a step which must inevitably lessen the respect felt for her, and violate the trust reposed in her by those who had loved and cherished her all her days.

"The marriage of girls of fourteen to seventeen years is a very prevalent cause of personal and transmitted evil and suffering. Prematurely taxed with the care and nourishment of children, their constitutions give way, and at thirty, they are already on the downhill of life. Eighteen is the youngest age at which any one should marry; twenty to twenty-three is much better."

But as it is, modern education, badly conceived and foolishly patronized, has but one aim, that of hastening as fast as possible, the bridal day. Mothers seem to rival each other in the hot haste with which they seek to secure suitable matches for their daughters. In this respect, their ardor is less excusable than the noble but mistaken instinct of a young girl, which leads her to an elopement, and certainly more injurious, because more universal.

A girl is hardly fit to receive attention in view

of a future marriage relation, before her eighteenth year, at which time her understanding and discrimination are fully awake, to guide the affections in a proper selection. She is then able to appreciate, understandingly, the character of those who approach her, and to love and esteem the true and manly one, who, by degrees, shall win her heart and hand.

In an important transaction like this, it is well not to be too much in haste. Persons who contemplate becoming partners for life, ought to understand each other's characters perfectly, before they solemnly consummate their union. For this purpose, the time from eighteen to twenty-one can be employed most advantageously. During this period, the affections of the young lady having been bestowed in love and devotion, should become settled and confirmed in understanding and esteem.

It is a misfortune of our times, that young ladies are allowed, and even hastened on, to *finish their education*, as the phrase goes. Its effect on the young mind is decidedly bad, as it produces the impression of having learned all that is needed, while in truth, we *never* finish our education; no one does, not even the most learned and accomplished. How, then, can we imagine, that a girl of sixteen or seventeen can have completed her education? Does she not need more knowledge than she has received during a few years in a boarding-school? Certainly; if she only would apply herself to acquire it.

But she has finished her education, and under this mistaken belief, many a young lady is thrown into the giddy whirlpool of modern society, into the lively, gay, yet oftentimes very monotonous circles of fashionable life. Here, indeed, more than in the quiet of home, will she be able to forget that she still has something to learn. So little of the truly needful and great in man does the present state of fashionable society require, that if the outward finish is only given, it pronounces *education to be finished.*

But we trust that sensible mothers and sensible girls will think differently from the leaders of fashion and its circles. They may participate in their gatherings, without becoming fascinated by their vain and transient pleasures, to such an extent as to draw them away from life's higher duties and more enduring joys. Hear a writer on education:

"When a young lady is seventeen years of age, if she enjoys good health, she is beginning to have that vigor of mind which enables her to make intellectual acquisitions. Two or three years, then devoted energetically to study, will store her mind with treasures more valuable to her than gold. She will be thus able to command a husband's respect and retain his love. Her children will feel that they have indeed a mother. Her home will be one worthy the name, where a mother's accomplished mind and a glowing heart will diffuse their heavenly influence. An angel might covet the mission which is assigned to a mother. Your child, who thinks of finishing her education at sixteen, may soon have entrusted to her keeping a son, in whose soul may glow the energies of Milton, or of Newton, or of Washington. God did not make her to play a waltz or dance a polka. She is created a little lower than the angels. When the waning stars expire, she is still to go careering on in immortality, till she reaches that happiness—in the presence of God. Appreciate the exaltation of her nature, her duties, and her destiny."

Education still remains the principal business of the young lady during the hours which she can spare from the performance of those duties she owes to her God, her parents, and society. How well she can employ these precious hours, as yet free of care and trouble; no tedium need overcome, no wish for excitement render her unhappy. Hundreds of legitimate branches of knowledge and accomplishments invite her attention, if she only is willing to follow the noble impulse of acquiring them, rather than of idling away the precious time of youth, or of spending it in vanities and gossiping. This is the period of life, emphatically her own, when more than at any other, she can dispose of herself, her time and occupations. This time she should, by all means, endeavor to improve; it is too valuable to be lost, for once squandered it never can be regained. How often have we heard the remark: If I only had paid attention to this or that in my youth, I would now know what to do; it is too late to learn it, etc. These confessions are humbling, but they are not the worst. Idleness in youth brings shame upon more mature years, and ought to be regarded as a foolish, unsafe companion.

But more than that, it is also criminal in view of our destiny as men, and our duty to provide in time for reverses which may come hereafter.

The poor have to be diligent: they have to acquire knowledge, in order to gain a subsistence; and the rich ought to do the same, in order to be useful to others, or to be able, if need should demand it, of supporting themselves by their own exertions.

We remind the reader of the many instructive lessons which have been practically taught, by those who experienced in exile, the horrible reverses of fortune during the French revolution Necessity then compelled many a delicate hand to work for a livelihood, either for herself or to sustain those she loved. Reverses of fortune, however, may occur at any time, and it becomes even the richest to provide for them. No provision excels the one which we may prepare within ourselves, by enlarging the sphere of our capabilities.

Young ladies may employ their time variously. Music, the arts and sciences, offer wide fields for pleasure and employment. Whatever is in this respect most agreeable to her, she has a right to choose. Besides the instructions in the household duties, which claim her paramount attention, she will find sufficient time to direct her mind to the higher spheres of knowledge and refinement. Let her bear in mind, however, that whatever she prefers in this line, she must have firmness enough to continue to cultivate with assiduity. One might prefer the study of mathematics, another one that of drawing, painting, botany, astronomy or the classics.

If she undertakes to cultivate any art or science, let her do so with heart and head in order to become perfect in it. In such a case only, it will be a source of great pleasure and recreation to her, refining the sentiments, enriching the understanding, and strengthening the will. Such is the education which we have to pursue constantly, and which never will end, not even with our lives.

A young lady, therefore, has not finished her education when she leaves boarding-school, but has just commenced it, and should continue it in such a manner that she may become perfect in some accomplishment, art or science, which would render her independent of the assistance of others, if adverse circumstances, or freaks of fortune, should overwhelm her and those near her. Fortune is the most fickle of dames.

Many of our readers may consider these allusions to mere external wealth and its uncertainty, as not belonging to the department of ladies, who in their sphere of life, have only the spending and not the making of money. It is true, woman in the beginning of society did not receive that special mission, to "earn thy bread by the sweat of thy brow," and during all subsequent ages, she has, as a general rule, been relieved from those particular cares assigned to man, as the provider of the family. But if this is and will remain true in regard to her destiny, the other sentence is not less so, that she shall be an help-meet to man, not merely to spend the means he has provided, but also to husband them wisely, and if possible and necessary, to enlarge them. It must be acknowledged that neither the education of our girls, nor the general tenor of the state of society, fit them for the fulfillment of such a task.

Neither can we hope that they will do so, while their parents indulge the morbid desires of emulating each other in making an external show of themselves, their houses, horses, carriages, furniture, etc., oftentimes to such an extent, that the complete ruin of a household and business reputation is the consequence. Those, whose yearly income would hardly, upon sober second thought, justify such aimless expenditures, nevertheless rush blindly ahead, under the mistaken idea of thus upholding their *respectability* in the eyes of those who happen to be Astors or Girards. Luxury and unnecessary expenses, to an incredible amount, have thus been augmented throughout the country, until they have become almost a part of its social existence.

From the non-producing class of society, those who by inheritance, or other favorable circumstance, became possessed of immense wealth, this love of external show in dress, houses, parties, etc., has descended to the producers, those who yet have to live by the sweat of their brow. It cannot be denied that the latter, in trying to imitate the former in luxurious appearances, have caused this lamentable state of things, this degrading and ruinous ambition of appearing wealthy. It is ruinous, because the experience of every day confirms it to be so; thousands of families every year sink down from a comfortable state of opulence, to penury and want, with mortifying remembrances of the past, and blasted hopes of the future. It is degrading, because it makes mere outward show, dry-goods and horses, a test of respectability, things and animals, the possession of which, should not, in the estimation of *sensible* people, add the *least* mite to one's station.

We here involuntarily remember the caustic anecdote of Franklin, related by him to the Legislature of Pennsylvania, that in case the right of voting should depend on the possession of property to the amount of twenty dollars, the death of an ass might deprive a freeman of that right. The illustration was to the point, and brought at once conviction to minds that shortly before were ready to vote themselves into the ridiculous position of granting to the ass the right of suffrage. Substitute, in his place, dry-goods, horses, carriages, and other such necessary articles of high life, wherewith some people fancy they become *respectable*, and they stand in the same ridiculous attitude; as soon as the horses are gone, their respectability departs with them. Would they might listen to Franklin's story, and follow its moral, as the Pennsylvania Legislature did.

But poverty and mental degradation are not the only injuries which result from expenditures inadequate to the income. The example thus held out by the inconsiderate parents to their children, leads them into every extravagance imaginable, and in the case of daughters, into a love of external finery and its concomitants, poisoning their very souls with falsities and dissipations of all kinds.

A young lady, once infected with this show-mania, will scarcely be able to fix her attention upon matters relating to the improvement of her mind. She has no relish for serious study, or the acquisition of those accomplishments which, if possessed in a certain degree of perfection, make her indeed respectable in the eyes of sensible people. She wastes the precious time of her youth in the most trifling and childish manner, preparing for the glitter of the ball-room, and listening to the shallow gossiping of parties In arranging her dress and outward appearance, she is not led to follow the dictates of her own sense of the beautiful, in doing which some good might result even from the folly of extravagance; no, she is continually dictated to by the representative of fashion, the milliner, or the Journal des Modes, which, like an inexorable tyrant, holds the whole female world in utter subjection. She has only to follow the stern mandates from Paris, and to please the Goddess of Fashion, has to make an exhibition of herself in the ball-room, theatre, at an evening, rather night party, or even in the street, where she often appears during the forenoon, in full dress, going from store to store, ostensibly on business, but really only to show herself in full, fashionable costume, to the passing crowds of a busy city, composed of all kinds of people, to whom, of course, her glittering appearance is a point of attraction. To be noticed by the crowd is the height of her ambition; the dignity of her female nature does not rise, to inquire into the real worth of this triumph of drygoods, fastened to her body. If it did, she would at once observe that she had paraded her finery mostly before a mixture of people, from the colored laborer in the street, to the stately senator on the side-walk. No one was prevented from feasting his eyes on the gaudy colors of her dress, and admiring the milky whiteness of her satin slippers, to say nothing about the costly display of sparkling jewelry.

In thus trifling away her best days, the young lady, brought up to such a life of frivolity and dissipation, cannot fulfill one of the most universal and beautiful laws by which the Creator has blessed the world, that of the bestowing of gifts for *useful* purposes.

No object or animated being is without its use in the world, certainly the glory of the creation, man, cannot form an exception. Animals follow their appetites and instincts, and their use is regulated by the necessities of nature. Man, on the contrary, fulfills his destiny by the dictates of reason, and the uses for which this power was given, are matters of his own choice, his own free will. This is a God-like position, and the consciousness of its relation is of such vast importance to each individual, that we cannot refrain from drawing the attention of the reader to its particular application in the case of young ladies.

What satisfaction can a rational being have in living without some laudable object worthy of pursuit, and the high destiny for which we were created?

We certainly must concede that pleasure in the form of gratifying the senses, by whatever means it may be, should not be the highest object of a rational being, either in youth or maturity. On the contrary, experience in a thousand cases, has taught us that a constant devotion to the search for mere pleasure, leads us away from the paths of duty and happiness, into the dark regions of despair and remorse.

It is true, we are all weak by nature, and prone to yield easily to the inviting voice of pleasure and dissipation. But if persons of a more advanced age and experience are not firm against the temptations of idle and mere pleasurable moments, how much more exposed to their follies and snares are those young in years and full of imagination and excitability.

Let a young lady commence her career in the right, truly humanitary direction; show her the purposes of life, the objects of her high destiny, and she will in all probability, continue in this path of usefulness and happiness. If this is not done, but in its stead, the young girl is encouraged by example and advice to commence her life in folly and fashionable dissipation, she will end it in misery and suffering. Hear Eliz. Blackwell, M. D., on this subject; she is an excellent observer, particularly in things referring to her own sex; her language is pertinent. In her admirable little book, " Laws of Life,"* she gives a faithful picture of the manner in which young ladies now-a-days, frequently waste their time, preparing the ruin of their future happiness. She says:

"The life of the young lady on leaving school, is little calculated to restore lost power, or to excite to a truer and healthier action by the presentation of noble objects of interest. What is there, in fact, presented to her worthy of pursuit? School discipline has not prepared her for serious study; indeed, study without an object, is of little worth, and she has no object in view for which grave preparation is necessary. The attention to domestic arrangements does not particularly interest her; indeed, by the age of sixteen or seventeen, she has learned all of household economy that she will learn at all, till called upon to practise it. There

---

*We recommend this excellent work to our readers, entreating them to give it a most earnest perusal.

are no schemes of organized benevolence to attract youthful activity and kind-heartedness, and teach to the young mind a deeper and darker lesson of life than it has yet learned; it is very seldom that the young can profitably engage in these enterprises. With the large mass of girls, gossip and frivolous amusements become, now, the chief business of the day; they have had no *serious* preparation for life, they know nothing of its *realities*, its *wants*, its *duties*—so the valuable moments are laughed and chatted away; every incident furnishes a theme for idle talk—church—society—promenades through the streets—all become subjects of gossip; novels are devoured to satisfy the new thoughts and desires that are springing up—parties, amusements of all kinds are eagerly sought for—the dictates of prudence, the requirements of bodily health, are alike disregarded; till at length, the giddy career is cut short by marriage!"

The above statement is not exaggerated, as every one will acknowledge, who has observed the occupations and doings of many young girls in our present day. One of the results of such a taskless career is, as rightly stated by Miss Blackwell, the early, premature marriage, hastily entered into without the knowledge of its duties and responsibilities. Parents are perhaps seldom aware, that they have aided in producing this unfortunate result by giving their daughter such ample opportunities to idleness and extravagance. The former leads her to seek excitement in novels, and in a society not much better than these useless and often bad books; the latter destroys all the relish for the cultivation of the serious objects of life and renders her a mere plaything for milliners, besides that it makes her indeed a *dear* child of her parents, who frequently wish her to be married that they may be rid of the expense.

Thus this early extravagance in dress and fashion often is the means of loosening the chords of affection between parent and child, becoming, thereby, the most awful enemy modern society has allowed to creep into its very life.

Notions of extravagance have been steadily on the increase ever since this country departed from the patriotic ways of its early fathers. In those times, true-hearted affection dwelt around the family hearth, and a father or a mother saw with tearful eyes and sorrowful hearts, the beloved daughter depart from the home of her childhood. She had been a daughter to them in the full sense of the word, never an expense or a trouble, and they felt in that moment her loss.

But now-a-days, the case is different; many a father feels relieved when the daughter's extravagances have to be paid by another one's purse, and in view of this, his paternal affection, which may be otherwise ever so keenly alive, becomes silent and stupified by the unnatural propensity of luxury and pleasure.

Thus, one evil creates the other; undue extravagance leads not merely to financial ruin, but often, also, to bankruptcy of soul and heart, the greatest calamity of the two, because the most irreparable. Nothing is so blighting in its effects upon individual and social happiness, as a morbid craving after the useless and giddy pleasures of so-called fashionable and high life. While a moderate participation in its pleasures is not objectionable, nay, becomes, sometimes, a recreation or amounts to a duty, as the case may be, an all-absorbing desire for them, annihilates the great purpose of human existence, and withers, particularly in young ladies, the flower of life before it can spread itself out in full bloom, to shed abroad the rich perfume of loveliness and usefulness.

To attain the latter end, that is, to prepare the young lady for her true destiny and duty, no efforts should be spared by parents or teachers.

Her physical constitution should be strengthened by sufficient and regular exercise, in-doors, by attending to the various duties of a house-hold and out-doors, by walking, riding on horseback, or other means.

Her moral and intellectual faculties should be cultivated by an earnest attention to, and study of, the various branches of art, science and religion. To interest her in the pursuit of such various knowledge, she should be allowed to bring each of them, when sufficiently acquired, into *practical* execution, as this, more than anything else, stimulates the female mind to their prosecution. The female mind loves less than the male, the abstruse and merely theoretical; whatever promises to be practical, engages her attention. She will, therefore, naturally shrink from the higher branches of mathematics, as too difficult or rare in application The same may be said of the dead languages, while the study of the living ones will be attractive to her, because she can make use of them at once as an accomplishment or in necessity. The acquisition of languages is particularly useful and appropriate for the female aiming at a higher and more refined culture of her mind.

Ralph Waldo Emerson very justly and pertinently remarks, that the acquisition of any new language doubles our existence, by opening to our perception and feeling a new world of ideas and expressions. To acquire three or four modern languages, may not be asking too much of a young lady who has time and inclination for the task. Her reward, after having accomplished it, will be immense.

French, German, Spanish and Italian, are those of the modern languages which chiefly recom mend themselves to her attention, from the fact that they are the most polished, and their practical application is the most ready and easy.

Of the natural sciences, botany seems to be the favorite of the female sex; no other science has such attractions, innate charms and facility for constant application. The knowledge and care of flowers and herbs, are naturally interesting and dear to every female, how much more if science gives precision and ease.

Of the arts, music and painting take the precedence; both are accomplishments of a high order and decidedly practical tendency. A knowledge of both is, therefore, desirable for a well educated young lady; they will add greatly to the fulfillment of many of the duties and purposes of her life, and repay an hundred-fold the time and exertions spent in their acquisition.

Besides the visiting of the sick and poor, a pious young lady will feel it her duty and pleasure to extend her benevolent designs beyond the limits of her personal presence. She will take an interest in the different great missionary efforts, which characterize so magnificently our present age; it is woman's influence which mainly supports them, and as she soon may be a wife and mother, she ought to feel the necessity to prepare herself early for such high duties and responsibilities.

CHAPTER. IV.

MAIDEN LADY.

The laws of nature, few as they are, form, nevertheless, a comprehensive and powerful conservative check upon the unnumbered multitudes of her creations, keeping them in almost constant subjection to her dictates, and thereby preventing many of the so-called freaks of nature. The law of development regulates each one's life from the cradle to the grave, and, willing or unwilling, each one has to pass through the different stages of life, subject in each, to various physical and mental changes, until mortality itself is stripped off.

The female infant will, in time, become a girl and young lady, whose physical nature is slowly but surely preparing to fulfill its part of the duties of a wife and mother. Thus a universal law of development hastens the female to a completion of her earthly destiny, and no contrivance on our part can change or delay the dictates of this imperious will. We are constantly in the power of this rapid stream of life, which bears us irresistibly to the close of our earthly existence. What we read in the ripple of its gently moved surface, or the surging of its threatening waves, are the duties of action, imposed upon us by each moment's advance, and the lessons imparted to us by close observation, reflecting beneficently upon our progress. Unable to stay the action of time on our bodies, which stealthily but surely follow nature's laws, we must content ourselves in each period of our existence, to have fulfilled its duties and learned its lessons.

Each maiden, therefore, should, in due course of time, become a wife and mother, this being the natural destiny of her existence. Her system, having developed all the functions necessary for that purpose, would find its surest integrity in carrying out the dictates of nature. In such a case, the harmony of the individual action, thus conforming to the natural law, would prevent the detrimental jarring otherwise inevitable and always pernicious to the welfare of the individual, physically and mentally. As a general rule, we may say that no female, having arrived at woman's maturity, should exclude herself from the duties of the married state, as well upon her own account as that of others.

But there exist exceptions to the most peremptory laws, those of nature not excepted; consequently we must look for them also in the present case. Young ladies, sometimes by predilection or force of circumstances, remain unmarried, preferring not to follow the laws of nature in this particular, at the risk of the penalties which their own chosen course might thus draw upon them. Generally a greater degree of health attends wives and mothers than those who remain unmarried; statistics also show conclusively, that they attain a more advanced age, less burdened with sickness in body and mind. These individual blessings nature seems to have intended as a compensation to those who, divesting themselves of all selfish motives, enter into matrimony, aiding, thereby, to sustain state and society.

This kind providential arrangement has been sufficiently powerful to produce the most desirable results, and the cases where females have voluntarily refused to marry, are quite rare, and then owing, perhaps, more to fanatical prejudices than real aversion. Persons eschewing matrimony on religious ground, will, of course, be the last to take medical advice on their intended step, and our reasoning with them as to its blessedness and legitimacy, would be all in vain. There are others, however, who, from different motives, choose to remain unmarried; to these we would address a few words of advice.

If a young lady of the proper age is healthy in body and mind, she ought to marry as soon as she has found the partner whom her heart prefers and her judgment approves. No trifling cause should be allowed to divert her from the path of duty and happiness which nature intended her to follow and enjoy. No petty selfishness, no fear of hardships or troubles should influence or agitate her mind to such an extent as to suppress the kindlier feelings of her heart, at a time of life when it is naturally the most capable of exciting and reciprocating love and devotion.

If this spring time of life passes by without forming an attachment from pure motives, the opportunity to do so hereafter diminishes in the same proportion as youth and the warmth of a young heart disappear. The heart does not calculate, but loves or hates; it has the most prominent voice in all those transactions of life where we have to choose or reject. The reasoning faculties only assist, not to make a choice, but to make a *rational* one. A young lady, if once her heart has chosen, should not, for trifling reasons, depose this heart, the queen of her existence, from the high position it is her natural prerogative to occupy. The weal or woe of a whole life depends upon this one decisive step; she can kindle and increase the spark of genuine love by disinterested and generous emotions, as well as by the opposite she can destroy it never to re-appear.

There are instances, however, where certain physical or mental conditions compel a young lady to refrain from marriage, although possessing a heart capable of appreciating and reciprocating the love of another. These ladies always are objects of the highest interest and the most sacred, respectful feeling for the lover of mankind. Entitled by nature, to the high privilege of bearing the Vestal flame, they are separated from the influence of mere earthly ties, and being clothed in the garments of purer loves, their presence sheds around them the glories and veneration of higher spheres. A heart thus sacrificed on the altar of nature, will not grow cold and feelingless. Although prevented from concentrating its rays of love on one single object, the sparks of genuine affection, once ignited in the generous bosom, rent in twain by disease or otherwise, will seek and find objects worthy of its love and devotion; an hundred virtues will adorn its holy, although seemingly solitary shrine; family and society will know them and bless, a thousand times, the heart which thus had suffered and thus had loved. Yes, such maiden ladies are a blessing to the family and the social circle.

They are generally found the most active and circumspect when sickness or distress invades the family or neighborhood. They minister most faithfully to the sick and render assistance frequently, under circumstances where other help neither could be had, nor would be equally as valuable. The good which thus they do often escapes the observation of those who reaped its immediate benefit; yet the Father of all good is also the Seer of all good, and fills their loving hearts with joy unspeakable. How often have we had occasion, during the performance of our professional duties, to see these angels of mercy, their hearts full of disinterested love, glide noiselessly from sick bed to sick bed, dispensing their kind attentions and good offices patiently and enduringly. Their presence would inspire the weary with fresh courage, and the suffering with new hope. Nothing escaped their attention, to render the chamber of agony and death as comfortable as circumstances would permit.

Monuments have been raised to the wholesale destroyers of human life, and at their feet nations have wept, bled and worshipped; but no one ever has found worthy of his praise or public esteem, the names or characters of the heroic women who, in the quiet walks of life, have risked their lives and fortunes an hundred times, without expecting reward or distinction. We will not comment on such glaring ingratitude, the weakest point of modern civilization. Do not let us boast of having attained, even a tolerable degree of civilization, before justice is done to the heroism of the heart. Let us honor the devotion of love and virtue, let us express gratitude and thankfulness to those who, by their sentiments and deeds, have redeemed and proclaimed the divine origin and sublime aspirations of mankind.

Considering the physical health of maiden ladies, it cannot be denied that their persistence in an unmarried state, becomes a source of many complaints and diseases, which, when married, might not befall them. Yet there exists one remedy which will counteract almost entirely the bad effects of impeded development, if it be only energetically and perseveringly employed. We allude to a sufficiency of physical activity, exercise, particularly in the open air, and the frequent use of cold water externally.

These hygienic means, in connection with a systematic performance of duties in and out of doors, will protect a maiden lady from those ills which an undeveloped system otherwise might engender, particularly if the body is too much at rest, and the mind irregularly and not fully occupied.

To accomplish these necessary objects, society offers a large field of action.

The family of which the maiden lady is a member, needs her assistance in the daily rounds of house duties; the mother and children lean upon her on every extra occasion, either in health or disease. In the parlor, kitchen and nursery, she soon becomes the *sine qua non*, who, if not present, is missed from either more than any other member of the family. There is no employment which can occupy more advantageously the attention of any female, than that which belongs to the fulfillment of household duties; and if the maiden lady's lot is cast within the range of a large family circle, she may indeed consider it a privilege to assist in supplying its various demands. It keeps the powers of mind and body active and elastic, preventing lassitude and depression of spirits.

The various nervous disorders so frequently met with in single ladies of middle age, who have generally led an inactive, sedentary life, will not appear in one who has fulfilled almost the same duties and assumed like cares with the mother of a family.

Besides, the family and society at large, can furnish her with a multitude of objects worthy her care and attention.

The church of which she is a member, offers several societies, religious and benevolent, which need her especial patronage and supervision. The different sewing circles for benevolent purposes, look to her for particular aid, and she ought to render to them as much assistance as possible. Schools of reform, which make their appearance now almost in every city, are objects worthy of her deepest interest; in the same rank stand the

Sabbath Schools, in which she ought to take a practical interest.

If a maiden lady has taste or inclines to cultivate the fine arts, she has objects enough to fill up her hours of recreation and amusement. Music and painting offer boundless fields of individual improvement and enjoyment, particularly suitable for the female mind. That which renders the arts and sciences such desirable companions of our leisure hours, lies not merely in their capability of refining our tastes and habits, but particularly in the unbounded resources they offer to the mind for that purpose. They must, however, not be made the sole objects of our activity, except by those who practice them professionally, else they consume all the personal attention, which fatigues and exhausts, without allowing us the inspiring and invigorating influence of their practical application as is the case with the professional artist. The latter, if deprived of the occasional stimulus of public praise or financial gain, would soon have his powers relaxed to such an extent, that he would be obliged to relinquish altogether, the prosecution of his studies. What we have presented here in the case of an artist, is a general law of nature, which operates in the same manner in any one's occupation. Our principal pursuit, or that which we make our business, needs, invariably, a stimulus of some sort, to neutralize the exhausting effect its continued prosecution would have on our body or mind. The truth of this remark any one can verify by observing, closely the different professions which compose society.

We need, however, besides our professional business, occupations which, by way of variety, become our pets, to occupy hours of leisure and recreation; these must remain what they are intended for, amusements and pleasantries, and as such, dare not be prosecuted with exhausting application.

Another great field of activity and profitable occupation for a maiden lady, consists in exploring the now almost boundless wealth of modern literature. After she has made the acquaintance of the various standard authors in her own language, she should find time to learn other languages, the acquisition of which, constitutes in itself, an employment of uncommon interest and benefit. Thus pursuing diligently the road upon which she has entered, that soon becomes a pleasure, which, in the beginning, seemed to be a hardship. The fibres of the mind gather strength from use in study and thought, as the muscles of the body from exercise. But as the latter is never more beneficial than when practised in the open air, so the former ought never to be prosecuted, save when having constantly all the doors and windows of the mind itself open to the free circulation of the literary air, which surrounds us. We need not retain any or favor all. This is the most glorious privilege of our free born nature, on the basis of which the most philosophic of all apostles, Paul, gave that splendid advice: "Prove all things, and hold fast that which is good." To keep the mind open and free for instruction, gives us the greatest facilities of applying the above rule, and also the opportunity of exercising the high functions of the mind, by rendering our judgment for or against. There is no better way of preventing bigotry and fanaticism, that mildew of the human heart, from taking possession of us, than by following the advice of the wise apostle, and thus making it possible to be both *active* and *useful*.

No position in life seems to promise a greater share of both activity and usefulness, than that of unmarried ladies, if they understand their mission in the family and in society; in both of which they form links as agreeable as indispensable. With a little care they will be able to preserve their physical health, at least from such ills as might result from the single life they have chosen. If such is the case, the lot of no human being is cast more favorably, and the future of no one shrouded in less darkness, because her responsibilities and duties, as they appear and are discharged from day to day, do not accumulate to require an account of her at any distant time. How different this in the case of a wife or mother, whose sphere of love and action, created within her own family circle, becomes so often the theatre of circumstances, which render it an intensely painful and not a pleasant arena. Although a mother often might enjoy pleasures more intensely felt, and perform severer duties more easily than any other one, yet none will deny, that just as often her heart will be lacerated with untold agony on account of those whom she loved so well.

Holier than a mother's love, and more remote from selfish interest, is the life of a maiden lady if spent in sincerity of motive and constancy of dutiful action. Of Christian virtues, none need be a stranger, but all may be gathered and treasured up within its range, and thus fulfill the demands made on every human being by the Holy One Himself.

CHAPTER V.

MARRIED LADY.

In contemplating the true position and destiny of a Christian wife and mother, her innumerable duties and responsibilities in the intricacies of social and domestic circles, of which she is the principal mover and ornament, we feel indeed, as if such a wide field of action and relation was too extensive to be discussed fully within the narrow limits of one chapter. Its importance and general interest would require a volume, in order to bring out all the detail belonging to this weighty subject. This has been done in a masterly style by Mr. Martin, in a work* entitled: "The Education of Mothers, or the Civilization of the World by Woman," in which the author conclusively shows, that this most desirable object *can only* be realized by the efforts of mothers, if themselves first rightly and humanistically instructed, that they may fully comprehend their

---

*This work was issued from the press of Messrs. Lea & Blanchard, of Philadelphia, several years ago, and merits to be in the possession of every mother, who considers herself to be a coworker in the great achievement, to civilize the world, which can only be accomplished by female influence.

duty, and be well prepared to overcome successfully the difficulties opposed to the realization of this grand undertaking.

Our purpose at present, consists in drawing the attention of wives and mothers to the importance of their own position, and in pointing out those physical and moral requisites, without which their high destiny cannot be fulfilled.

The wife, in assuming that name full of meaning and power, has given, as the phrase goes, to the husband of her choice, her heart and hand, in fact, all that is claimed to be worthy as a possession. At first sight, therefore, she has voluntarily renounced will and power, the two faculties by whose exercise everything around us is governed; apparently nothing is left to her, to influence society or the world at large, and consequently her existence might be considered to be without weight or power in the family or social circle. Yet, this same wife and consequent mother, shall be its redeemer and powerful protector; history and experience teaches that it is so. The question naturally arises then, how can this weak creature, who has divested herself of all rights and self-disposition, become the powerful, all-governing influence in the world?

This seeming paradox finds its solution in the fact, that a married woman, although having ceased to dispose of herself only, being, as it were, in the power of her husband, nevertheless, disposes not alone of herself, but also of those thus closely related to her. This she effects by enthroning herself in the affections of the husband, absorbing them entirely, and thus becoming, as it were, the soul of the union, while he is its head. It is well known, however, that where the affections are, there our influence preponderates in readiness and weight of action. The intellect, as represented by the head of the family, acts merely as the balance-wheel in nearly all transactions of daily life, and thus far the husband exerts an influence, regulating, as it seems best, the propelling force of inclination and will. The latter altogether centres in the wife, and this immense power, the female desire under male direction, can be wielded by her for good or evil. In either case the results are proportionate to the wide sphere of action, which comprises the whole of human society, and the transcendant importance of those interests, depending upon domestic and social relations.

It is all-important, therefore, that woman should exert this, her legitimate influence, in a manner to insure the greatest amount of happiness all around, and the possibility to retain the power thus acquired and directed by her. She is the soul and mistress of her husband's affections; her wishes become his own, and if approved by his judgment, are speedily carried into effect. Through him, therefore, her influence has to manifest itself; for and through him she lives; without him she would have lost the centre and strength of her actions. To him she looks in distress and adversity for relief; he is her only and constant confidant; no secret of her heart is kept from his knowledge; in joy and sorrow, he is the partaker of the heavings of her swelling or suffering heart. His gain or loss is her own, his plans or operations interest no one so much as the loving partner of his bosom.

Thus the constant change of affection is the real talisman of the marriage union; and it is the wife who keeps this precious jewel, whose mysterious workings unlock a paradise full of love and happiness. Woe to her, however, if she should not possess it; and still worse, if she should lose it, after once having tasted the sweets of its presence! Nothing can compare with the misery into which both parties would then be thrown, because no earthly treasure could buy or supplant the absence of that love which descends from higher spheres only into the human soul. But there is hardly a human being in existence who has not received a spark of this heavenly flame, sufficient to kindle affection in hearts where otherwise it would have slept unlit. The frequent exhibition, therefore, of love betrayed and affection unreciprocated, is not caused by total absence in either of the parties, but by the improper method of showing them.

Affection is the daughter of love, whom she never precedes, but always follows. If love departs, or is not to be seen, affection cannot appear; when a wife ceases to love, or neglects to show her love to her husband, no affection will be kindled in the bosom of the latter, no harmonious exchange of thought and feeling can exist. Love manifests itself particularly in a close attachment to its object; a wife must cling to her husband, who is the centre of her life's orbit, like the earth, who rolls her axis in restless joy around the sun, her centre of life; if she would depart only for a short time from her orbit, the rays of the sun would fall cold and cheerless upon her joyless surface; she would not be able to distribute innumerable blessings, nor even shine herself with light of her own.

This comparison has not merely the merit of a close analogy, but contains the expression of a law of nature, which pervades in equal force her moral as well as physical domain. Its further application, as to marriage relation, we leave to the contemplation of the reader, who will find on closer examination of this subject, the truth of our remarks.

Not external beauty or splendid attire, is the

sole attraction a married woman should offer to her husband; they are gifts and acquisitions not to be despised by those who possess them, but never important or enduring long enough to be envied by those who chance not to have them. If present, they should form only the casket in which to preserve nobler qualities of the heart, which never lose their charms, or fail to produce affection and devotion in the manly breast.

It is needless here to enumerate the many virtues which should adorn the character of a wife; they are abundantly known and understood; besides, we have already grouped them together in the early chapters of this book. We may be allowed, however, to draw the attention of the reader to a few of the most important ones, although sometimes the least observed, on account of their seeming insignificance.

Cleanliness in all that pertains to the wife's domain, is an indispensable companion of her virtues, which are reflected in its spotless surface, as in a mirror of infinite value. To a tidy and well arranged home, the husband hurries his steps with a more than ordinary longing of the heart, and his affection for its beloved mistress is already warmed into life, before he meets her on the threshold of his comfortable dwelling. Here she greets him with the confiding devotion with which she is attached to the heart of her choice. She may not be able always to smile, because circumstances may make it impossible. Nevertheless, in fortune or adversity, she rushes into his presence with an unfeigned expression of relief and security, which shows him at once his duty and pleasure. Because attachment and its expression on the part of the wife, form the great load-stone which moves the affections of the husband, and always draws her the nearest to his heart.

While the wife is thus adorning herself with the most magnificent jewels which can enrich any female character, she ought not to be unmindful of the duty and rightful policy of attending carefully to her external appearance before her husband as well as others. Neatness and taste in dress and attire, more than a gaudy or showy appearance, will most favorably engage the attention of the pleased and admiring eye, and the husband, as any other one, has received eyes which would like to behold objects of beauty and symmetry, if presented to his view anywhere; how much more, if in the person of his wife, already beloved and revered, he can behold and admire these pleasing qualities of form and figure.

It is true, the pleasure here portrayed, is one in which the senses, from the first at least, are more engaged than the mind. Nevertheless, its reaction upon the inner man is not the less sure and effective; **all our likes and dislikes generally** originate in the same manner; the various impressions of external objects on our senses form the greatest part of our joys or pains. There is hardly an exception to this rule; the wife can the least expect to form one, as regards her husband, because none has a greater right to enjoy and admire the beauty and symmetry of her form or figure. If the husband is pleased and delighted, her purpose in this respect is fully obtained; she needs not cater for the admiration of any other.

Neatness of dress, like piety, that incomparable quality of the soul, requires a general, uniform, and constant exhibition, else it cannot sustain itself long, or improve indeed, the tastes of its possessor. If a wife is neat only at certain times, in the presence of her husband, while at others, she appears in a slovenly attire, the contrast thus created is certain to neutralize the benefit of her first effort. The impression on the husband's affections must be decidedly unfavorable. Neatness, moreover, is well calculated to preserve the dignity of the female sex, certainly better than seclusion and haughtiness, which, in accomplishing this end, destroy other precious qualities of the soul, equally as desirable.

A neatly dressed wife will less often forget her lady-like nature, and rarely sink below its standard, while one appearing in a loose and careless garb, will hardly make an attempt to elevate her feelings above the lower promptings of our nature, allowing her soul's attire to assume a character similar to the slovenly appearance of her garments, and thus the intercourse between husband and wife will often become less courteous or dignified, a misfortune for both greater than at first sight it would appear.

There lies great danger in these slight beginnings of relaxed attention or interest, both of which are so essential in keeping the matrimonial flame alive. Let a wife once lower herself in the estimation of her husband, as regards neatness of appearance or sweetness of temper, and she has thereby opened a Pandora box, full of misfortune and unhappiness. Step by step, but steadily, the social and familiar conduct between the two loses in interest and purity, and gains in lightness and frivolity, until it is depraved to an extent which leaves nothing but moral and physical ruin, where once a bright future opened its inviting portal.

It is not our province here, to go into the minutiæ of married life, prescribing certain rules to be observed by the wife or husband, in order to insure the greatest amount of happiness during the matrimonial career. Nothing is farther from our purpose than this; we abhor a censorious, dictatorial spirit, but wish most heartily that every one, capable of forming a judgment, should exercise this high prerogative of human nature, to cultivate all his faculties in such a manner and degree as best befits himself and his own welfare. What we desire to do is, to indicate the leading features of a conduct, which, by their general import, control all the minor ones belonging more strictly to individual character. Variety in unity, is the greatest charm of nature in all her creations, and why should we not allow a variety of conduct between man and wife, if the principal requisites are but observed to keep harmony?

Charity covers a multitude of sins, and no one needs the constant exhibition of this exalted attribute of a noble soul more than a wife in the daily intercourse with her husband. No human being is perfect, or will be so this side of the grave; it would be, therefore, unreasonable in a wife to require of her husband perfection. And since he cannot be without some imperfection, it is her natural duty to judge his conduct with that charitable and loving disposition, which allows her to be of real benefit to him, by inducing him to alter it as much as possible. Whoever forgives the most and with the greatest readiness, is said to be the best Christian. This truth ought to find the widest application in married life; no state in which we can exist here on earth, affording a greater opportunity for its exercise; in none can it be brought to such practical perfection.

We will now proceed to contemplate the various

physical changes which take place in married women. They are generally productive of good, tending, if not hastened by too early a marriage, to strengthen and consolidate the constitution, notwithstanding the apparent hardships and sufferings which a woman has to undergo during pregnancy or parturition.

The immediate effect of marriage on the well and fully developed female organism, consists in the greater vigor and increase of vitality and sensibility in all the organs influenced by the physical and moral changes which have taken place in her condition. She is now married; her highest ambition and most fervid wishes are fulfilled. This state of her mind reflects sensibly on her nervous system and circulation, and increases their action beyond former limits. She is lively, gay and sprightly, and her whole system partakes in this holiday feeling of the senses, developing itself more fully and densely in all parts. The enjoyment of the senses, so cautiously approached, but at the same time so rationally bestowed by matrimony, opens a new portal in the secret chambers of nature for another series of physical developments, different from what had taken place in the organism of the girl, during its preparation for womanhood, but based upon these proceedings, which are now brought to a higher and final perfection.

Menstruation, as the paramount feature of girlhood, now loses its exclusive importance, and vanishes for a time entirely from the physical theatre as soon as conception takes place.

The reader will recollect that a fully developed girl is only so on account of the complete development of the ovaries, whose periodical activity shows itself in the formation of an ovule, and its displacement from the ovarium through the Fallopian tubes to the uterus, whence it is carried away abortive with the menstrual blood; the latter discharge having been caused by the same productive stimulus which congested the ovarium. When in matrimony, however, the conditions of nature are fulfilled, this ovule becomes impregnated with the vital force, which enables it to change at once the process of those organs which formed, and now shall develop it, in other words, shall harbor it for a certain time and make it grow.

Before we consider any further the uterine life of this new being, let us delineate those signs by which its presence in the womb can be ascertained almost from its earliest existence.

There are but very few general symptoms constant enough to be reliable in all cases, which indicate conception and pregnancy in its earliest stage. We will name them here, however, in order to enable the reader to form a judgment of them, if they should occur. She feels a singular emotion of painful pleasure, and a shuddering proceeding from the spine; a pain in the region of the navel, sometimes a sensation of motion in the abdomen, and a tickling in the region of the hips; she feels fatigued and sleepy; this state is followed by a sense of fullness, warmth and heaviness in the abdomen.

The first more certain sign of pregnancy, is the suppression of menstruation, which, if not caused by other morbid circumstances, indicates that the internal surface of the uterus, from which the menstrual blood is secreted, is now engaged in other secretions, stimulated into existence by the presence of a fecundated ovum, which has been retained in the uterus, adhering, generally, within its upper portions. Sometimes menstruation may continue for several months; in such cases the menstrual blood is secreted from the lower portions of the uterus, while in its upper, the changes take place necessary for the growth of the fœtus.

Another well attested sign is sickness at the stomach, with which a great many women are troubled in the beginning of pregnancy; it is uncertain, however, in as far as other congestive states of the uterus, such as suppressions of the menses, etc., may produce it, without having the slightest reference to pregnancy.

If the above signs are present in consequence of true pregnancy, other symptoms will soon develop themselves to verify it. These are enlargement of the breasts, brown circles around the nipples, appearance of milk in the breasts, and finally an enlargement of the hypogastric region. All these signs found together, form a plausible array of symptoms in favor of the existence of pregnancy, yet they are in themselves not sufficient to prove it positively, because they may be produced by other morbid agencies. Women, who have had children, possess signs, which individually, are mostly sure in indicating pregnancy. For instance, some have always tooth-ache, styes on the eye-lids, or black spots on the face, neck or hands, like freckles; others are taken at once, without a known cause, with salivation, which in some instances proceeds for a long time, unless mitigated by the use of ale, champaign, or Scotch herring; still others have strange desires or longings, modifications of the appetite for unusual substances, such as chalk, etc. All these signs, belonging strictly to idiosyncrasies, are, therefore, no absolute signs of pregnancy, but only accidental.

The most sure sign of pregnancy which a woman can have, is the quickening, or the motion of the child. It generally takes place at four and a half months from the beginning of gestation, and serves, therefore, as a mark of reckoning, being the middle of the time allowed generally for the duration of pregnancy. In but very few cases, quickening occurs either sooner or later. When this takes place, it may be safely presumed, in connection with the former signs, that a true pregnancy exists, which has, at that time, already run half its course.

A fecundated ovule, if it shall grow, must adhere to the sides of the uterus; if it does, we may consider conception to have taken place in reality. Then the uterus is stimulated to secrete from its walls on all sides, a membrane, called the decidua, lining the whole internal cavity of the uterus, and forming the medium between fœtus and mother, by which the former can come into communication with the blood circulation of the latter. Within this membrane a circulation is established, which unites the child with the mother, through the cord and the after-birth, the former adhering to the child, the latter to the uterus, in which it roots, like a tree in the ground, both containing veins and arteries for the flux and reflux of the blood. The blood of the child does not go over into the circulation of the mother, to become decarbonized, but is oxygenized by being exposed in the finest ramifications of the placenta to the oxygen carried thither by the arteries of the uterus. Thus the great purpose of oxygenizing the blood is carried on without the necessity of inflating the lungs with atmospheric air, which, of course, would be impossible in fœtal existence.

The fœtus, or young being in the womb, from this time up, grows and develops itself with astonishing rapidity, in the thousands of intricate parts which constitute the human organism. Though bound together by the vital force in one harmonious whole, the various parts of the different systems develop not all simultaneously, but gradually crystallize, as it were, into one whole body. This formative process consumes more than one-half of the uterine life of the new being. If no disturbing influences interfere, it will develop in a perfect manner; but if morbid causes should operate on the fœtus, its harmonious development may be intercepted, and its growth arrested at any period during gestation, in certain parts of the system, while others develop themselves naturally. This fact explains the origin of those organic imperfections and deformities which characterize the so called monsters, whose singular appearance is sometimes attributed by the ignorant to mysterious causes. To this class of arrested fœtal development belong also most of those cases, where children are born with marks on their bodies or limbs, not developed or even entirely wanting. It is not as yet sufficiently settled, whether such a state of things can be produced through the influence of the mother on the child, some physiologists denying its possibility, as no nervous connection between the two is as yet found to exist. Whether or not fear, fright, etc., operating violently on the nervous system of the mother, can affect the child, we know at least, that misfortunes of this kind are best averted by avoiding those violent nervous emotions.

Bodily exercise, or even fatigue, is easier borne by a pregnant woman than mental excitement. While the former increases her physical health, and consequently that of the child, the latter disables the energies of her system, and must injuriously reflect on the child's development.

It is the duty of the future mother to live and act during gestation in a manner that her offspring may be benefited thereby. Nothing must be omitted to realize this, the principal object of her life. She must expect, beforehand, to undergo all kinds of hardships and to make severe personal sacrifices; the thought that it is for the benefit of her future offspring, will strengthen her to an indefinite degree of fortitude and courage.

She must take daily exercise, sufficient and of an active nature. For instance, riding in a carriage does not belong to the active exercises; walking, particularly in the open air, is more active, and therefore better calculated to invigorate the system of a pregnant woman. The best, however, is that exercise which accompanies the daily attendance to the various duties of a household. In these the pregnant woman can find employment sufficient for body and mind up to the very last hour of her confinement, and more suitable to invigorate her own and the child's health, than in any other. She must not merely direct the affairs of the house; such a course would not accomplish the ends in view. But she must be active and busy herself, if it is only in the lighter kinds of work, such as setting the table, sweeping and dusting the rooms, etc. If she does not feel well at such a time, particularly if she is threatened with abortion, she must not commence, or if she has, must cease at once, to exercise in the above manner. A perfectly healthy woman, however, should not neglect these exercises under the mere pretext of being unbecoming or too fatiguing.

Next to exercise, the most important consideration is dress, which ought to be as loose as possible, in order to allow the most ample liberty for enlargement. Corsets or tight clothes generally, are very injurious, as every one will readily concede. It must be understood, however, that she ought to dress sufficiently warm at all times, to feel comfortable. If she has been in the habit to wash and bathe in cold water, the state of pregnancy, even far advanced, does not prevent her from following this most important hygienic rule as long as it is convenient or shows its beneficial effects by re-acting on the system in a healthful glow.

As regards the various disorders during pregnancy, we intend to treat of them in the second part of this book, where the reader can find them under their proper heads. Here we would state, that an otherwise healthy woman ought to pass through the whole of her pregnancy without any considerable feeling of disease, and this, too, until the very commencement of labor pains. That this is the real intention of nature, becomes apparent in the thousands of instances where women in the pregnant state, do feel neither disease nor are incapable in body or mind, to attend to all their accustomed duties, enjoying life as much as ever. That this is often not the case, is owing to the irregularities of life and departures from nature's dictates, by which women become weak and an easy prey to disorders. We remind our readers, therefore, of the great rule, the more natural or less artificial you live before and during pregnancy, the less sickness and discomfort you will have to encounter during pregnancy. Let strict hygienic rules guide your life and but very little ailment will disturb your ease and comfort.

Quickening is a term by which is generally expressed the first perception a woman has of the child's muscular action, and not as if life, at that time, first entered the child. It may vary as to time and power of expression; in some women it occurs earlier and stronger, because the child may be more active and strong; in others it is retarded and weak, because the child may be less lively and powerful or the quantity of the child's water greater. Upon the whole, however, it may be said that quickening occurs in the middle of pregnancy or four and a half months from either the conception or the birth of the child. Yet, very many and great exceptions to this rule exist; in some women it may never manifest itself, in others it may appear as early as two months and a half after conception, as one lady I knew myself used to experience. Syncope or fainting is not an unfrequent accompaniment of quickening, but soon ceases after the woman has become more habituated to the peculiar feelings of motion in her womb made by the child.

About this time the womb has extended so much as to raise itself out of the pelvic cavity, thus acquiring more room for extension and the facility to rest on its brim. From this time up to the seventh month the growth of the child is very rapid and perceptible in the outward appearance of the future mother, the most prominent feature of which is the so-called *"pouting out of the navel"* or its protrusion, which takes place between the sixth and seventh month. The navel instead of forming a hollow, is now pressed outwardly by the force of the gravid uterus which sometimes causes the navel to protrude.

In the seventh month the child has acquired such a perfection of development in all its parts, that it is able to subsist outside the womb, if accident or disease should have hastened its birth. Any birth previous to the seventh month is, therefore, called very properly an abortion, indicating, thereby, the impossibility of the birth of a living child, although some exceptions may even here take place; while a birth at the seventh and before the ninth month is designated as a premature birth, because the child *can* live, although prematurely born, and consequently of difficult raising.

Pregnancy has, in most cases, a duration of nine months, each of thirty days, although sometimes it lasts two hundred and eighty days, or ten months, each of twenty-eight days; cases have even occurred within my knowledge, where ladies were not confined until the three hundredth day after conception, which constitutes ten solar months, each of thirty days. Accordingly we see that a variation of thirty days is possible, or that a regular birth may occur between two hundred and seventy and three hundred days of pregnancy.

The appearance of the so-called labor-pains soon terminates the state of gestation, expels the child, together with the after-birth, and allows the womb to contract, thus assuming, by degrees, its natural size again. Although this process is a physiological one, which, as such, should not meet, on our part with much interference, yet the pain and distress to be borne by the woman, are sometimes so great as to make their mitigation desirable This fortunately can be effected to a certain extent by the use of ether, which being poured on a loose handkerchief, is inhaled by the sufferer Chloroform, although more decisive and certain in its anaesthetic action, is not, on that account, preferable to ether, because its action being too violent, endangers the life of the patient, even in those where no diseases of the heart or lungs can be presumed to exist. We warn our readers against the use of chloroform in an emergency of any kind; while we can recommend the ether from practical experience, as useful and harmless. It should not be applied, however, until the severe pains of the latter stage call its use into requisition, when its effect need not be such a protracted one, which ought to be always avoided as much as possible.

The duties of a mother, perhaps the most responsible and severe imposed on a human being by nature, seem to be presaged and presented to her mind before even the child is born. The future mother prepares, beforehand, what is needed for her offspring when it appears. In this respect man partakes of the same instinctive providence which has been bestowed by nature upon all animal creation. The birds make their nests for the comfort of their young, and man provides carefully for the child's comforts long before it needs them.

But as soon as it has appeared in this world of pain and pleasure, the realities of a mother's duties never cease to present themselves to her. They may vary in kind, but never cease to exist; moreover they grow in intensity of interest and responsibility as the child develops itself from earliest infancy to maturity. A mother remains always one to the child of her bosom. The father may finally cease to exert a father's influence over his child; his parental relation may change into that of a friend's. A mother never changes her character as regards her offspring; she remains the same to him that she was from the beginning of his existence, the careful guardian of his physical and moral welfare. Time and advancing age cannot work a change in her feelings towards her child; she remains faithful to the trust imposed upon her by nature. In her the conservative element prevails to a greater extent than in man; and if the latter can even so far divorce himself from the innate parental feeling as to disown his child, and slay it even as was the case in the Roman Senator, who broke the staff over his own son's life, the mother never yields to other voices than those of nature, or other dictates than those of the heart. Wonderful provision of the Creator, who thus made the otherwise weak, the instrument of protection and preservation, while the strong, impelled by moral force, is not unfrequently disarmed into leniency by the melting power of natural affection!

Watch a faithful mother in the discharge of her duties in the nursery, at the side of the cradle, or amidst a group of boys and girls of different ages; observe closely, how she manages to get along. Her tact in this respect is wonderful, beyond the grasp of man's most exalted intellect; no rules of art can supply its want. The talent of ruling judiciously in the charming world of children, is a peculiar gift of nature, bestowed upon woman alone, as her sole prerogative, which in faithful hands becomes truly enviable. Then we perceive the wonderful versatility of woman's mind; now she coaxes into quiet slumber the restless eye or whining mood of the infant, now she reproves the noisy restlessness of the boy, and satisfies in stately teachings the eager curiosity of the girl. But always is her eye watching over the whole group, catching at a glance their physical and moral wants; and the expression of her language or actions is suited to the sex or age of the child. In vain could such admirable conduct be asked of a man under similar circumstances; his patience would give out, before even his mind could comprehend the task; hopeless would be his situation, if he had to fulfill similar duties. If he has to act on the large stage of life his part, she has to do the same on a smaller, but not less complicated one; neither would fill satisfactorily the place of the other. And if this is true, as it certainly is, the conclusion is irresistible, that each one should remain within the sphere of action properly assigned to each by the ordinance of nature. Let the father be indeed a provider and protector; let the mother be indeed a guardian and teacher of the little ones comprising the family.

Fashionable life, we know, has sadly disarranged this beautiful order of nature, and thereby weakened, in many a respect, the ties which bind the filial tree to parental roots. Yet we think, that its laws are so deeply seated in the soil of human feeling and society, that it is almost impossible to destroy their rule or annihilate their existence. Even fashionable life, with its many heartless and unnatural traits, cannot divest the most reluctant and giddy mother of the inward pressure of her soul, to witness the frolicsome up roar of a nursery-room, or to spend hours of delight and care in the company of her little ones.

Nature designed the mother to be the child's first and most legitimate teacher; from her lips the infant perceives the first sounds of the sweet language of love and caresses; to her the narrow, but nevertheless ardent wishes of its heart are first directed and look for their realization. As the child draws the nourishment with impatient delight from the mother's bosom, itself heaving with pleasure, so does its longing eye rest on her countenance, radiant with intelligence, and beaming forth the tender affections of a mother's love into the appreciative soul of her offspring. Where is enjoyment like this to be found in the whole range of nature's economy? Here she has gathered a combination of physical and intellectual wealth, seldom to be equalled, and never to be excelled, and has laid it at the feet of that being, whom above all she desires to enrich with her choicest blessings, in order to compensate, as it were, for the extraordinary charge laid upon a mother's shoulders. No triumph of the victorious soldier, the subtle statesman, the eloquent advocate, successful physician, or faithful minister, can equal in reality and sublimity a mother's happiness, when fondling her infant.

No other avocation in life can give the full reward in innocent pleasure and joyous feeling. There is no reason, then, why the woman, with true motherly feelings, should engage in any other but the business assigned to her by nature, for the transactions of which she is so eminently and exclusively qualified, and so richly rewarded. It is true, she often also experiences sorrows and sufferings, which no human being can feel with equal intensity. The sickness of her children disturbs her rest and grieves her soul. But these very misfortunes carry within themselves a higher stimulus to a mother's action than the mere every-day duties could afford, and like the passing thunderstorm on a sultry summer day, leave life's atmosphere clearer and more beauteous than ever. If death, notwithstanding, calls away from her motherly heart one of the beloved ones, she follows its flight into the regions where hope and faith combine to promise a future re-union. Thus, even the worst misfortune which can befall her, the death of her offspring, enriches the domain of her love, giving her an undying interest in the joyous realms of heaven, which thus become more than ever the point of her pilgrimage.

No member of a family has a greater interest in the welfare of society than the mother. Although the child of the past and destined to direct the present, the mother feels anxious to shape the future of that society in which her children have to play their part, either for good or evil. This one thought alone is powerful enough to arouse her energies to the utmost, and engage them constantly for the physical and moral welfare of society. She will, therefore, not allow herself to be a stranger to the political events of her country, because these, touching the dearest interest of all families, must affect her own also.

And here may I be allowed to pay my humble

but sincere tribute to the lovely and heaven-born genius of womanliness, whose spirit always supports the right, and persists in doing so heroically, until the cause has triumphed. Our present age, so conspicuous in reforms, is mainly indebted to woman for those which have a tendency to ameliorate the condition of the poor and suffering. It is woman's peculiar privilege to follow the appeals made to the heart. A cause, therefore, which has within it the germs to arouse that delicate womanly sympathy, will soon enlist it in its favor, and be successful. We mention in support of this fact, the different benevolent and missionary societies, ragged schools, schools of reform, orphan asylums, temperance societies, etc. All these various institutions owe their principal sustenance to the charity of woman, and mostly to the mothers of the country, who feel it besides to be their duty to enlist in this moral crusade for the recovery not of the Lord's grave, but of the Lord's temple, filled with the holiness of Christian conduct. This is the field on which woman can show the heroism of her soul; if here she is not to be seen, engaged in the battle against evil, it is plain she has not understood her mission, or has forgotten her most sacred duty. If our modern times can be proud of a thousand distinguished features, possessed by no former age, it is not merely because we have steam and hot-air power, railroads, and electric telegraphs, which mostly influence the material welfare of mankind, but because we are engaged in reforming the world morally through those means, of which we have above named only a few. These comprise together, what the Scripture calls "the sword of the spirit," and under its action woman's flag has enlisted. The greater part of the glory, therefore, of having civilized the world, and prepared the way for its redemption, falls upon woman; a fact which becomes more and more understood, as the work of Christian reform progresses and shows its magnificent results. Wherever good shall be promoted in the world, there woman is found ready to lend her helping hand. If King Alcohol has to be dethroned, women fill the halls of temperance, listen to addresses, and sign petitions; if the atmosphere of a prison shall be cleared of its foul moral miasm, woman is at hand to do it; if hospitals need attendance, woman is willing to go there, frequently at the peril of her own life. Oh, let woman but persist one century longer in those beneficent efforts she has begun of late so vigorously, and soon the earth will be a paradise again, prepared to be the habitation of the Holy One! Mothers, your mission is a glorious one.

CHAPTER VI.

WIDOW.

The last duty to a beloved husband and honored citizen is performed. Slowly the crowd disperse to their own homes, variously impressed with the solemnities of the occasion. The religious among them, while witnessing the departure of an immortal soul for judgment and eternity, are busily engaged within their secret chambers of thought, to review their past, making resolves according to the degree of faith they severally enjoy. The worldling might have been seen trembling with uncertainty while hearing the sound of the falling clods on the coffin, which in a similar manner he now feels, will at one time receive him. The thoughtless among the fashionable crowd, gaze with pallid but stupid faces, at the circumstantial ceremonies which solemnize the day of mourning. All, however, soon forget the hour and occasion, which thus has called them together, and return to their various predilections and avocations, as if the picture of death and immortality had no lasting effect on their thoughts or actions.

But not thus ends this day with one, who in that crowd might have been seen bowed down with grief and sorrow, unable to observe either her own or the state of others. Still, while the excitement lasted, accompanying such distressful circumstances, while the chamber of death was full of kind friends and neighbors, and the avenue leading to his final resting place, evinced in the thronging crowd, the love and esteem which the beloved dead had thus far even carried with him, her strength remained equal to the weight of sorrow on her heart, because not as yet did she fully realize that she was alone; as yet did she breathe an atmosphere made congenial by a husband's presence, even if in the shrouds of death. Reluctantly she returns to the home, from whence the sun of her life has just been removed, and where now, for the first time, she experiences all the depressing feelings of loneliness and mental solitude. Thus, for weeks, months, and even longer, does the widowed heart mourn for its departed object, turning its inward emotions towards the past, where it was surrounded with the love-inspiring affections of a devoted friend. No proof can be stronger of the fact that man was created to reach his highest perfection on earth in *matrimonial* union, than the perseverance of this attachment, beyond even earthly limits, by the removal of one of the partners, which we frequently witness, aye, in our times, where matrimonial choice is not always the most genuine.

It seems that the memory of the departed, particularly the remembrance of his many virtues, which often live in his deeds after him, and surround his widow with a glory as lasting as the name she bears, becomes, by degrees, the means to fill up the vacuum in the widow's heart, and sustain her in future trials in such a degree, that she never feels the need or want of another attachment. Then the great object of life is fulfilled; such a heart remains still married. Its sun has not disappeared entirely; he has only withdrawn his enlivening rays beyond the theatre perceptible to earthly senses, but still accessible to those of the heart, where they are felt as intensely as ever, creating reminiscenses of by-gone times, with the power and sweetness of present realities.

If external circumstances do not too palpably remind such a widow of the loss of her husband, if she is beyond temporal want, and the care of her children an easy task to her, then indeed is her situation not without joy or real pleasure. She soon will habituate herself to the cares of the day, and while she remembers the manner in which her husband used to overcome them in his day, strive to revive his memory, by daily imitating his example. Thus widows have frequently continued to fulfill the most arduous duties of business and station for years after the death of their husbands, showing a tact and perseverance equal to that of their former partners, and proving clearly how elastic and formative the mind and will of woman is, if compelled or set free by the force of circumstances.

But we must be well reminded, that a woman performing, in such a case, a man's duties, does or can only do so, without losing her feminine character, by having become the heiress of her husband's mind as well as estate. In his spirit, while it still is with her and rests upon her, she continues to conduct affairs, which, under other circumstances, would seldom be inviting enough to draw her activity away from her own legitimate female sphere

While thus pursuing her course, she represents a double nature, exhibiting both strength of will and tenderness of heart, a combination of perfection so seldom found in this our world of faults, and in the widow's case, the result of suffering and attachment, which invest her with an uncommon degree of sympathy and interest.

But not every widow can be expected to be so constituted, or placed in such circumstances, that she can do without manly counsel or assistance. Then the broken reed will be lifted up again by a new power. The elasticity of her female nature will rebound by the impulse of the new force; her heart will swell again with new emotions, and a fresh love, sometimes equal in fervor to the first, will rejuvenate her whole being, and elicit sparks of affection again in the manly breast of a new partner.

There is nothing unnatural or forced in the formation of a second marriage by a widow, under circumstances which open her heart and direct her wishes again towards the world around her. If this is the case, such a marriage is sanctioned as well by the customs of society as dictated by the necessities of the laws of nature, assuming thereby all the legitimate and hallowed attributes of a first love. Nature, in her richness, has strewed her gifts in exuberant profusion, and it would be strange indeed, if in doing so she had forgotten to endow the poor human heart, which has to feel so deeply, and suffer so intensely, with a faculty of re-invigoration, in order to lift it up again with a new courage of existence, and to double its days of youth and joy. None must question sincerity in actions, where nature and law has granted the right to act.

A widow will find her greatest consolation in the education of her children, endeavoring at all times to supply in their training, the place of him who, leaving them fatherless, trusted them in her care. Although nature, generally, did not endow woman with a sufficient determination of will and force of character, to eminently qualify her to rule the heedless boy into willing obedience, and on that account, it might seem at first sight impossible for a widow to train her older sons into respectful behavior, yet such is the power of sympathy created in every bosom for the widow's defenceless position, that soon even the wildest boy will feel within him such promptings of a nobler nature, as will impel him to respect the one whom he sees alone and without a protector in the struggling world. If the mother can inculcate into the young mind of her son sentiments of this kind, at an early time, exciting his sympathy and emulation, she has won his affections already, she has conquered his passions before they arose, and made him tame before he was wild. She must commence early to revive in his memory the image of his departed father; his many virtues, and the love and respect he bore to her and his children. The memory of such a father will always be dear to the tender emotions of a child, and turn the undue increase of passionate excitement into love and reverence. How often have we seen the most stubborn boy corrected and made submissive by the simple address of his widowed mother: O, John, what would father say if he was yet living, and could see you acting thus. What in this respect, a living parent is sometimes unable to accomplish, the timely remembrance of a departed one, of his love and virtue, frequently has happily effected. In this manner, the revered shade of a beloved husband still hovers around the lonely widow, protecting her with a more sublime, because more spiritual power than when on earth. To his image and character the helpless widow directs the revolting spirits of her bolder sons, and the fury of those young lions is soon tamed; to his memory she appeals, when in the struggles of a heartless world, it becomes necessary to defend herself from mercenary aggressions, and in most cases the appeal will be successful.

A widow, therefore, true to her interest and calling, will cultivate the memory of her departed husband with delight and care; he is not dead to her, whose heart contains no other image but his, whose soul still swells with the sweetest emotions, even at the bare mention of his name. Years may pass away; with them does not pass away or diminish in brilliancy, the picture of one who was all in all to her while living. A widow in that sense of the word, has still a husband, a protector, a counselor, a guide.

As regards her physical welfare, she has, in view of her widowed position, no other rules to follow but those general hygienic laws, which it is the duty of every one to obey, in order to preserve health. Her system has been fully developed during her married state, and is, therefore, not more prone to diseases, than others who are still married. Activity of body and mind, variety in both, and cheerful contentment, are essential in all conditions to health and happiness.

CHAPTER VII.

MATRON.

When the evening of life draws nearer, and the sun of its physical appearance inclines towards the verge of the horizon, then the lamp of the inner man begins to shine brighter, illuminating with more than earthly lustre, all the recesses of a heart, tried so often and trained so well, and now ready to forgive and to forget everything, but the love it bears to every one around, and which never was so holy, because never so disinterested as now.

The wife and mother has become a matron and grand-mother; a great change has taken place. Her physical condition cannot favorably compare with that of former years. Step by step, her senses weaken; her eye-sight grows more dim, her hearing more obtuse; she has lost the elastic movement of her limbs, and the ready eagerness of mental pursuit. Slowly but surely, all earthly ties seem to loosen, and none seem finally to be left, strong enough to fasten her flying spirit to interests below. Yet one feeling never forsakes her, one thought always is present in her mind, which already almost lives in other spheres. The same love which filled her heart in younger days, and made her endure and suffer for the sake of others so heroically, still glows within her, and keeps alive the interest between the outward world and her retiring spirit. Love, that mysterious child of heaven, is still the silent but powerful telegraphic messenger, which unseen, but not unfelt, travels patiently over the wires of the soul, from the heart of the aged matron to the objects of its longings, which she is daily leaving farther behind her. No physical decay can reach this immortal attribute of man, and prevent its manifestation, as long as life and consciousness remain.

A matron, in her relations to family and society, is invested with a peculiar interest, and excites our most lively attention. Removed by physical inability, from the common strife of the day, yet fondly attached to remembrances of the past, which revive more vividly than ever, she becomes, by comparing them with the present condition of society, an able critic and fearless monitor. She cherishes the young, and re-kindles on the fullness of their joyous vitality her own expiring years; the innocence of early years seems to re-enter the place within her heart where it once dwelt, and happified her whole existence. The experience of a whole life, spent in arduous pursuit and discharge of duties, seems only to have been the means of preparing her for the re-union with qualities of the soul, belonging preeminently to the days of inexperienced childhood, where no passionate excitement as yet disturbs the placid waters of contentment. She is contented, she is delighted in the company of children; like them, she has no wish for gain or ambition of distinction; like them, she enjoys the moment in the innocent pleasures it bestows, be they ever so insignificant in the eyes of others. Frequently she directs with evident relish, the various sports of the little ones, being careful not to offend their tender sensibilities by any unkind remark or action.

More judicious than a mother, and more patient than a father, the matron rules the child-world more satisfactorily than both. Because she herself, delights in what she creates for their amusement, and a child soon distinguishes between those who really sympathise with its whims and joys, and those who only affect to do so. She is, therefore, not neglected by the younger generation, but rather preferred to others nearer to them in age and relation. This fact alone constitutes a moral triumph of no mean importance, as in the economy of nature, it becomes a counterpoise to the physical inabilities and ills of old age not to be attained by any other means. Cheerless, indeed, must be the winter of life, if not enlivened by the jingling bells of infantile chattering, and the restless motion of childhood.

In company with adults, the matron appears in all the dignity which age and ripened experience confer. The deference and esteem paid to her in the social circles, at home or abroad, are well earned testimonials of her worth, and sit lightly upon the character of one who is conscious that in every relation of life she was anxious to do her duty to others from none but disinterested motives. It is a satisfaction for the matron of a house, to be consulted in matters of importance, because the exercise of judgment in more advanced age becomes a necessity of the mind's activity, as in younger days activity itself forms a part of life's nourishment. Blessed is the son or daughter, who can call upon an aged mother for counsel and advice in times of need, or matters of importance. The wisdom of higher years has improved by age, as the wine does in its quality; unlike the judgment of younger persons, it is free from the unrefined, poignant acerbity of selfishness, which yet adheres to those of little experience in worldly affairs. A matron's useful lesson and cautious warning saves them, therefore, many a failure and annoyance, even in the common business transactions of life. Thus, the knowledge resulting from a life of labor and exertion, is not entirely lost or spent in vain; it is transmitted to posterity through daughter or son, and becomes the means of still greater development. We cannot describe here all the relations of life in family or society, where a matron's prudent counsel exerts the most beneficial influence; they are too numerous and varied.

But not the mind alone should be kept active in the highest stage of life, in order to lighten the burden of an existence, which drawing towards its close, becomes physically more and more sluggish, but the body also should enjoy as much exercise as is possible, and compatible with comfort. It is too often the case, that aged ladies allow themselves to indulge in the habit of too much and long continued sitting. This is particularly injurious, when they incline to corpulency, as it begets a tendency to apoplexy, or hastens its occurrence, if the tendency to this disease already exists.

Daily exercise, plenty of fresh air, and the free use of water externally, is as necessary in this advanced stage of life, as in former ones; a difference, however, exists, as regards the free use of water internally, which in old age is not so much called for, nor so beneficial as in young persons. A little good old wine, at or after dinner, is more apt to aid the weakened digestion of old people, than the use of cold water, because old wine contains just that stimulus which suits the diminished re-action in a declining system. The wine, therefore, is justly called by the ancients, "*lac senum,*" the milk of the aged, and if used only at dinner, in a small quantity, produces vigorous digestion, sustaining thereby the strength of the system.

Elderly persons, although enjoying a good appetite and healthy digestion, should nevertheless avoid eating to satiation. It will be better to eat often, but little at a time; in this respect, the same rule obtains in old age as in childhood. Changes of temperature should be strictly regarded, as they frequently produce the most dangerous consequences in old age. Exposures to cold and damp weather, are particularly injurious to the lungs, and cold wet feet to the abdominal organs, which are at best the weakest part in an aged system.

Almost all diseases which may befall women at this stage of life, have no characteristic peculiar to the sex, and we must, therefore, refer the reader as to their treatment, to the physician, or to those works treating on this subject in general.

RECAPITULATION.

Thus, have we traced the female constitution and character, from its first appearance in the girl, developing through all the stages and changes of womanhood, as wife and mother, to the last period of its existence, the old lady, or matron. We have seen this female character, under the most varied conditions of life, still evince such a harmony of expression with the physical constitution of woman, that we cannot for a moment resist the conviction that the latter is the basis of the former. In arriving, therefore, at the true destiny of woman, we were compelled to begin by interrogating her physical nature. In doing so we soon perceived the limits of her career, not merely in misty outlines, ill-defined and easily to be overlooked, but, on the contrary, well-defined and visible to any one whose intention is rather to observe nature than to criticise her institutions. These critics of nature and her laws have existed in all times, and their voices have been heard in all directions, expressing dissatisfaction with the rule of the laws of nature, and seeking to substitute in their stead laws of their own make. How sadly the world would fare, however, under the government of these hypercritical philosophers, a single incident in the life of one of them might demonstrate. While walking in the shade of some stately oak, our philosopher, being well pleased with the grand appearance of this king of the forests, observed on close examination that its fruit, the acorn, did not correspond in size with the magnificent proportions of the tree, and accused nature, or nature's God, of a great error in not having allowed the oak to bring forth fruit of the size of a melon, which would correspond so much better with its giant proportions. Having thus exercised his critical acumen, and being satisfied as to the righteous judgment in these matters, he laid down in the grateful shade of the criticised work of nature, and fell asleep. During his sleep, the wind agitating the branches of the oak, caused one of the acorns to fall, and, as it happened, on the nose of our philosopher, who awaking in terror and amazement at finding his nose bleeding, exclaimed: "Fool that I was, to correct nature's laws! If a melon had thus struck me instead of an acorn, I now would not live to see my error, and adore the wisdom of the Creator!"

In observing the part woman has to play on the great theatre of life, we have found how just and economical everything in her own and the condition of her fellow-actors is arranged, to bring about the greatest effects from the smallest means; how the weakest being, as woman appears to be, can and does execute the greatest and most comprehensive work of life; how on her depends the welfare and glory of all; and how she only can fulfill this wonderful mission by either intuitively following the promptings of her nature, or by reading and studying closely the features of her destiny, as they are unmistakably laid down in her own constitution and the relation it bears to man and society. The sphere of woman is the most exalted in the whole domain of nature; let her but fulfill the mission assigned to her, and let her perfect but her own legitimate career, and the world will soon everywhere, as it has already been done in our own blessed country, acknowledge her right to rule society, which it was her sole prerogative to create by forming the family and its sweet sociability.

To the above we will add a stronger testimony and more urgent appeal in the very comprehensive language of M. Aime Martin. "There is no universal power on earth, except that of women. Nature has given to them the superintendence of our childhood, and the control of our youth. As children we owe them our thoughts; as young men we lavish our sentiments upon them; and they preserve at a later period as wives, that influence they had acquired as mothers and as mistresses. Thus the entire circle of our life rolls on beneath their influence. *The mission of weakness is to regulate strength; the mission of love is to make us delight in virtue.*

"Oh, woman! could you but have a glimpse of some of the wonders promised to maternal influence, with what a noble pride would you enter upon this career, which nature has generously opened to you during so many ages! That which is not in the power of any monarch, of any nation, it is sufficient that you should will it in order to execute it. You only upon earth dispose of the generation which is just born, and you alone can re-unite its scattered members, and impart to them the same impulse. That which I could only write upon this insensible paper, you can engrave on the heart of a whole people! Ah! when I see in our promenades and public gardens this boisterous crowd of little children who are playing around, my heart beats with joy from thinking that they still belong to you. Let each of you labor only for the happiness of your child; in each individual happiness God has placed the promise of the general happiness. Young girls, young wives, young mothers you hold the sceptre; in your souls much more than in the laws of legislators, now repose the futurity of our nation, the world and the destinies of the human race."

PART II.

DISEASES

OF

WOMEN;

THEIR DESCRIPTION, AND HOMŒOPATHIC TREATMENT.

CHAPTER I.—Diseases of Sexual Development.
" II.— " of Generative Organs.
" III.— " of Nervous Function.

CHAPTER I.

DISEASES OF SEXUAL DEVELOPMENT.

I. PUBERTY AND ITS ABNORMAL APPEARANCE.

In the first part of this work we have explained the meaning of the term *puberty*, when applied to the female system, and have mentioned the various processes and phenomena through which it has to pass and manifest its presence. If this development of the system occurs in a normal order and degree, no disease accompanies its manifestation, and we behold the girl transformed into a woman, approaching all her characteristics, without the slightest degree of sickness or distress. Not always, however, does this great change, which has its principal seat in the ovaries, take place in such regular and healthy manner. It may appear too early or too tardily; the development may be an imperfect one, or may not commence at all; or if it has made its appearance, the various phenomena may not occur in harmony with each other. In all these cases it is evident that the whole organism must participate in the morbid movements of the sexual development, and create disorders which are in more or less intimate connection with it. To this class of diseases belongs what is called

CHLOROSIS.—GREEN SICKNESS.

Diagnosis.—The patients have a peculiar color of the skin, not excessively white, as we see it after severe loss of blood, but a paleness with an admixture of yellow and green; the lips appear at times almost white, the lower eye-lids swell, and appear darkish blue; the skin, rather loose and flabby, feels cold to the touch; the patients themselves cannot bear a low temperature, and wish to be where it is warm. The tongue shows an unusual pale color, and is frequently covered with a thick, tough mucus. The patients evince great muscular weakness; tire very soon after slight exertions; love, therefore, rest, being apparently lazy. The same languor expresses itself in the operations of the mind, which is listless and without energy. The patients breathe hurriedly, not, however, because they have a difficulty in breathing, but on account of not having muscular strength enough to take a deep breath; an examination of the lungs would show no morbid alteration of the texture; the heart palpitates considerably, particularly when going up stairs, and the pulse is accelerated, sometimes to 140 beats in a minute, yet not full, but small, weak, wiry, and easily to be compressed. The veins of the skin appear of a pale rose color, never distended as in health; and the blood in them is thin and watery. With a diminished appetite, the patients have a slimy taste in the mouth, pressure in the pit of the stomach, and cructation of wind after eating, even of the most digestible nourishment; sometimes there is an immoderate desire for eating chalk, charcoal, etc. Digestion is deranged, causing sometimes excessive constipation, followed by a diarrhæa of substances badly digested. If the affection proceeds unchecked, the lower extremities become swollen, hectic cough sets in, sometimes with expectoration of dark-colored clots of blood, symptoms which have all the appearance of a rapid decline. In some cases the nervous system becomes sympathetically affected, producing hysterical fits, spasms, even somnambulism.

We have given above a full description of this disease, to enable the reader to recognize it from its first appearance. Although the disease is evidently the result of an inharmonious development of the sexual functions, yet we cannot positively fix the cause upon one particular function in all cases. Sometimes menstruation had not made its appearance before the disease sets in, and we naturally infer that its suppression has produced it. In other cases, however, the disease can develop itself, though the menstruation has appeared, but then it must have been either too early or too profuse; which circumstance, reflecting deleteriously on the simultaneous development in other organs of the system, produces chlorosis, by deteriorating nutrition. Any want of harmony, therefore, in the development of the sexual function, can excite this disease. As regards climate, the northern may retard too much, and the southern may too greatly accelerate the formation of the menses. The same may be said in regard to the conditions of life; in the poorer classes every thing has a tendency to weaken; in the richer, to over-stimulate the constitution; either of which has deleterious effects on the normal development of sexual functions.

More immediate, exciting causes of this disease are those extremes of the mind, exhibiting either a love of indulging in frivolous phantasies and immoral connections, or a depression of spirits, a melancholy caused by home-sickness, troubles of all kinds, and particularly by disappointments in love. Marshy regions, damp dwellings, excessive exertions or sedentary habits, immense loss of blood, in fact, everything which can weaken the constitution, while it needs all its strength to develop the sexual functions, can produce chlorosis.

Treatment.—As this disease, on account of its great importance, always needs the attendance of a skillful physician, we merely intend to draw the attention of the reader to the principal remedies to be used in the beginning, the application of which frequently prevents its progress. The cause which may have excited it must be removed, if possible, before an application of medicine can be of much avail. If chlorosis occurs in girls, when the first menses have not yet appeared, the treatment mentioned for obstructions of menstruation, in the following chapter, will be suitable.

Among the remedies *pulsatilla* will be best adapted to females of a mild disposition, given to sadness and tears, or if exposure to cold or dampness was the cause; if there is difficulty of breathing after slight exertion, sallow complexion, alternating with redness and flushes of heat, palpitation of the heart, cold hands and feet, looseness of the bowels and leucorrhea, cough with expectoration of dark coagulated blood, mental and physical langor. *Sulphur* after the above remedy, if the patient is not relieved. *Bryonia* alternates well with *pulsatilla*, if there are frequent congestions to the head, bleeding at the nose, dry cough, bitter taste in the mouth, and chilliness, with pain in the small of the back. After *sulphur* it is frequently necessary to give *calcarea carb.*, if the oppression of the chest is very great, and the extremities begin to swell, after which *ferrum* should be given in repeated doses, particularly when the sallow hue of the face continues with great debility, want of appetite, nausea and hectic cough. The above medicines should be given in long intervals, say every third or fourth day a dose, (six glob.,) until amendment takes place, or another remedy is indicated; *ferrum* may be given in the first trituration, the other remedies in higher potencies. If this disease occurs after severe sickness or hemorrhages, give *china* and *carb. veg.*, every other evening a dose, (six glob.,) alternately for at least five or six weeks. In both cases, if the above medicines do not relieve, apply to a skillful physician without delay.

Application of Water.—The frequent use of the sitting-bath in the morning, and the sponge-bath in the evening, is very beneficial; during the night the patient should apply the wet bandage around the abdomen; but during menstruation all application of water must be omitted.

Diet and Regimen.—Let the diet be very nutritious, and make the patient exercise freely in the open air; if mineral springs are chosen for a summer resort, give the preference to the chalybeate, containing iron.

MENSTRUATION AND ITS ABNORMAL APPEARANCE.

Of all the phenomena connected with the sexual development of the female, menstruation claims our greatest attention; although, as we have seen in the first part of this book, it is the result, rather than the cause of this development. Yet it is one of its most prominent signs, and we can be greatly guided by its appearance in our actions, if such are necessary by the presence of disease. Menstruation may show its abnormal character in various ways, either as *too tardy* in its appearance, or *suppressed* after its appearance, *too copious* or *too scanty; difficult* and *painful;* it may return *too quickly* or last *too long*, and finally it may appear on *other places* than the natural one in the system. We will now consider the above points successively.

Tardy Menstruation.— It is of the utmost importance first to know whether a girl, although old enough to be menstruated, is developed sufficiently otherwise to make it necessary for the menses to appear. As we have seen in the chapter on girlhood, menstruation is the result of changes which, at the time of puberty, have to take place in the ovaries, and without which no discharge will be possible, and if forced by medicine to appear, will destroy rather than establish health. As long as the girl has not increased in size across the hips, or the breasts have not become fuller, indeed until the whole form and conduct of the girl shows that this change has taken place in the internal parts, no effort must be made to force nature, as it would be not only fruitless, but really injurious. In such cases a general treatment should be instituted; the girl should be made to exercise freely in the open air, and not be allowed to frequent school, if in the habit of attending. Nature will soon rally her formative forces, and herself establish without force what medicine could not do without injury. But if the girl is fully developed otherwise, (see p. 87,) if she suffers from time to time or at regular monthly intervals from pains and congestions either in the head, breast, or abdomen, it is necessary for us to interfere by giving some of the medicines, as described below. In such a case also it is indispensable to regulate her habits, and make her follow the most strict hygienic rules. She ought not to study or sit long at a time, but rather exercise in the open air, either by walking or riding in all kinds of weather, although she must not be neglectful as to the changes of temperature, taking especial care to have her feet always dry and warm. And here we may be allowed to remark, that in the course of a long practice we have found the wearing of thin-soled shoes to be the most fruitful source of the decay of female beauty and the decline of female health. The injury of tight lacing (although considerable) is nothing in comparison with the fatal habit of appearing, in all kinds of weather, in thin-soled shoes and light stockings; the consequences of the latter are beyond description fearful and destructive. She should avoid all **highly seasoned food and stimulating drinks,** such as coffee, tea, wines and malt liquors; water, however, externally and internally should be used frequently and freely.

In a majority of cases, besides prescribing the observance of the above hygienic rules, we have found it sufficient to administer to the girl, in order to establish the first menses, *pulsatilla*, every evening a dose (four glob.) for four evenings, then to discontinue for four evenings, after which to give *sulphur* in the same manner, if necessary. This treatment we have generally commenced a few days before the return of the monthly distress could be expected. Just at this time a warm foot-bath may be administered in the evening when going to bed.

If these remedies are not effective, we must have recourse to one or more of the following remedies:

Bryonia, if the girl looks flushed in the face, her nose bleeds frequently, and she inclines to constipation.

Veratrum, if she is chilly, has cold hands and feet, palpitation of the heart and disposition to diarrhea. *Bellad.*, if she is frequently troubled with congestion to the brain, redness of the eyes, intolerance of light, giddiness when stooping, having a full, bounding pulse, yet cold feet. In such a case *aconite* should be given alternately with *belladonna*. If, however, spasms should threaten or indeed break out, with nausea and vomiting, screaming, even convulsions, *cuprum* should be given in alternation with *belladonna*, very half hour a dose, (four glob..) until better.

Phosphorus is indicated, when the breast and lungs seem to suffer in females of a delicate constitution, weak chest, but lively disposition; particularly when, instead of menstruation, an expectoration of small quantities of blood occurs with a hacking cough and pains in the chest.

Arsenicum is particularly useful, if dropsical symptoms appear, swelling round the eyes in the morning, and of the feet or ankles in the evening, besides a feeling, as if hot water was coursing in the veins.

Suppressed menstruation, or its *temporary* cessation, when once having been well established. This occurs either suddenly, by taking cold, (from wet feet mostly,) over-heating, violent mental emotions, faults of diet, in consequence of other diseases, such as rheumatism, or diseases of the lungs, liver, uterus, etc., in which cases it often produces violent congestions of the chest, head and stomach, with cramps, convulsions, inflammations and with various other symptoms of disease; or the menses have disappeared, without creating any complaints; in which case it may be caused by pregnancy, or lay the foundation of serious chronic disorders, not at that time apparent.

If the suppression of the menses has been caused by the presence of other diseases in the system, such as rheumatism, etc., as above referred to, the cure of these disorders has to be effected by an appropriate treatment, before the return of the menses can be expected.

If, however, the suppression is a sudden one in consequence of other morbific causes, producing violent symptoms, select from among the following remedies the most homœopathic, and administer it either in solution (twelve glob.) in half teacupful of water) every hour or two hours a teaspoonful, or in the dry state (four glob. in the same intervals until better, or until another remedy is indicated.

Pulsatilla is here also the principal remedy, particularly in females of a mild disposition, inclined to weep readily and feel melancholy; if the suppression was the result of getting cold or wet; if the patient suffers from headache, especially on one side, with shooting pains in face, ears and teeth; palpitation of the heart, suffocating feeling, flashes of heat, though the face looks pale, nausea, vomiting, pressure in the lower part of the abdomen, disposition to diarrhea, frequent desire to void urine.

Opium, if all the blood seems to have rushed to the head, producing *heaviness* there with dark redness of the face and drowsiness, even **convulsive jerks of the muscles.**

Bryonia, either after or in alternation with *pulsatilla*, especially if there is great tendency of the blood to the head, producing a swimming, heaviness and pressure, particularly in the forehead, worse by stooping or moving; congestion and pains in the chest; bleeding at the nose, dry, hacking cough, with pains in the lungs, like pleurisy, pains in the small of the back.

Aconite should be given first in all cases where the congestion of head or chest is very great, pulse full and bounding; throbbing, shooting pains in in the head, sometimes producing delirium. If fright is the cause, its application should be followed, if necessary by *opium*, especially if the head is very much congested; and by *veratrum*, if severe cramps in the abdomen are present.

Belladonna in alternation with *aconite*, if in young and plethoric persons the congestion of the head is very great, accompanied with dread of light, fever, burning thirst, bearing down in the lower parts of the abdomen towards the genitals, red and bloated face.

Chamomile, when the suppression was caused by a severe fit of anger; *ignatia*, if caused by grief, fright, chagrin; *opium*, if from sudden joy or fear: *hyoscyamus*, if from unhappy love or jealousy, particularly when accompanied by convulsions, sudden starts or involuntary laughter during sleep, desire to run off; *mercury*, if caused by home-sickness, and then alternately with *phosphoric acid*, if the patient is dull, listless, the hair falls out, has profuse night sweats and hectic fever.

Sulphur in all cases where *pulsatilla* has failed to restore the catamenia, or where the suppression did not excite very violent symptoms immediately, but threatens to lay the foundation to chronic disorders. In such cases it must not be oftener given than once in eight days (six glob.)

Too Copious Menstruation—Flooding.—If the discharge is too profuse or lasts too long, it is not composed any more of real menstrual blood, which, being a secretion from the uterus, is different in quality from the blood in the vessels; but it is common blood, passing away from the womb, either in paroxysms or steady dropping. Such cases are generally attended by pains in the back, loins and abdomen, resembling labor pains. They are either caused by fright, fear or other mental excitements, or external injuries, such as a fall or blow; ulcerations in the womb also produce them. If mental causes have preceded, the reader will find the same remedies beneficial as related above for suppressed menstruation, especially if caused by mental agitation.

If external injuries produced the flooding, the use of *arnica* internally is necessary. In every case, however, complete rest in a horizontal position is indispensable, the head being not higher than the hips; the covering may be suited to the feelings of the patient; it generally ought to be more cool than warm.

If other than the above causes have produced the hemorrhage, we must select from among the following remedies.

Ipecac., in alternation with *china*, is the principal remedy, particularly if there is great weakness, buzzing in the ears, faintness, when raising the head off the pillow; the blood is bright red. *Belladonna*, if there is a *bearing down* sensation. *Pulsatilla* and *lachesis*, if the flooding occurs during *change of life*, at which time *ipecac.* and *secale* are of great benefit, particularly in decrepid women.

Chamomile, if dark, clotted blood is discharged, accompanied by colic-like labor-pains, violent thirst, coldness of the extremities, headache, with clouded sight and humming in the ears.

Crocus, if the blood is dark colored, clotted and very copious, and the menses have appeared before the usual time.

Platina, in the same case, particularly if attended with bearing down pains, great nervousness, sleeplessness and constipation; alternates well with *belladonna* and *chamomile*.

Coffea and *camphor*, in alternation, when there is, beside the above symptoms, *exceedingly* painful colic.

Secale, if the discharge occurs in elderly women,

attended with great weakness and coldness of extremities, and then in alternation with *china*, if the discharge has been copious, or with *carb. veg.*, if the weakness from the same cause is very great. In such cases small quantities of old wine, such as Madeira, but frequently given in some water, are of great service.

If real *flooding* ensues, resisting the above medicines, the application of *cold water* or *pounded ice* over the lower part of the abdomen, externally, becomes necessary, in order to coagulate the blood in the vagina and uterus, by which means the hemorrhage generally ceases. There is no danger of getting cold in doing this, if it is only done right; the patient must be lightly, but well covered, in a position strictly horizontal, and as quiet as possible. This treatment will be effectual in the severest cases of flooding. If cold water is applied, the bandages must be renewed as often as they become a little warm, say every two or three minutes; they must be wrung out well at each renewal.

At the same time, when this external process is going on, the proper internal remedies should be continued. They must be administered in solution; six globules, of a remedy, in six tea-spoonfulls of water, every fifteen, twenty, or thirty minutes a tea-spoonfull; the intervals should be extended as the patient gets better.

The drink given to such patients must be cold and not of a stimulating nature; except when faintness appears, with deadly paleness of the face, no pulse, and cloudiness of sight, give wine and brandy in frequent but small quantities. *Camphor* and *china* are, in such cases, of the greatest benefit.

Menstruation *of too long duration* or returning *too quickly* requires the use of the same remedies as above described under menstruation too *copious*. If it occurs at the period of *change of life*, compare the article on this subject.

Menstruation *too late* and *too scanty*, require, in the main, all those remedies enumerated under the head of menstruation *too tardy*. *Pulsatilla* is the principal remedy, when the menses appear too late or do not discharge enough, also when they are irregular, sometimes too late and too profuse, (which occurs mostly at the critical period or change of life,) sometimes too early and too scanty. In the former *lachesis* alternates well with *pulsatilla*. In most all cases *sulphur* is necessary to a complete cure.

Deviation of menses is a term signifying the appearance of a monthly discharge of blood from other places of the system, such as the lungs, bowels, nose, stomach, etc., which has the effect, that while it lasts, the real menstruation cannot appear. This aberration, as it may properly be called, of a discharge so vitally connected with the sexual functions, has been observed to have taken place from all organs and parts of the system, bowels, stomach, rectum, fauces, nose, gums, urinary organs, respiratory organs, eyes, ears, ulcers and wounds, wherever they were; the principal places, however, are the nose, stomach, lungs and the end of the rectum. As regards the causes of this singular phenomenon, a great diversity of opinion exists, as yet, among the authors; we have mentioned its occurrence here, to enable the reader to recognize its appearance, and to administer such medicines in its beginning, as are advisable. Should the disease resist those remedies, the advice of a skillful physician must be sought at once.

If the discharge appears monthly through the *eyes*, *chamom.*, *nux vom.*, *carb. veg.* and *bella.* will prove beneficial; if through the *nose* or *ears*, *bryonia*, *mercury*, *rhus* and *silicea*; if through the *lips* and *gums*, *bryonia*, *mercury* and *phosphorus*; if through the *fauces* and *lungs*, *phosphorus* and *bryonia*; if through the *stomach* by vomiting, *bryonia*, *carb. veg.*, *veratrum*; if through the *bowels* and *rectum*, *nux vom.*, *arsenic*, *sulphur*. The latter remedy will be the most important in every case at the end of the cure to prevent a relapse.

Too difficult, *painful* menstruation, *menstrual colics*, are of frequent occurrence, and constitute ailments of no slight importance to the health and comfort of the person thus afflicted. Their causes generally can be traced back to the early years of womanhood, to improper treatment of other diseases, suppression of eruptions or habitual discharges, rheumatic disorders, colds, etc. The pains either appear before or during the flow of the menses, sometimes resembling real labor pains, with bearing down and forcing; at other times there is a constant aching in the loins, hips and limbs. They generally diminish in violence as soon as the regular flow commences, but not always. We found the following treatment in the majority of cases the most successful.

As soon as the patient feels the pain commencing, she should lie down, cover well and take *coffea, pulsatilla* and *veratrum* in alternation, every half hour a dose (four glob.) *until better.* If this does not relieve within two hours, take *nux vom.* if the *forcing* pains predominate; *cocculus* if *colic* pains appear in the abdomen, with shortness of breath; *chamomile* if with discharge of dark colored blood, there are pains resembling labor pains, together with colic pains and tenderness of the abdomen.

If a profuse perspiration sets in while in bed, the patient must not leave it immediately after the pains cease, nor cool off too quickly, else the pains return. Avoid the use of heating substances internally and externally, save a hot brick on the the feet or stomach.

CESSATION OF THE MENSES OR CHANGE OF LIFE.

This period, commencing about the age of forty-five, forms one of the most important in the life of a female. If not guided through this critical time by the counsel and aid of a skillful physician, she gathers the seeds of endless misery or even early death. This period may be a blessing to her as well as a source of great distress; because after it her health becomes either more confirmed, or disorganizations in internal organs may form with fatal consequences. Without enlarging here further on the subject, we recommend the early and constant advice of a skillful physician during this period, which generally lasts from one to two years.

Change of life begins with an irregularity of the menses; they cease for two, three or four months, thus giving rise to a suspected pregnancy; then re-appear with great violence, then cease again for four or six months, during which time the woman shows more or less symptoms of congestion; piles appear; the limbs swell; pruritus (violent itching of the private parts,) sets in sometimes; also cramps and colics in the abdomen; asthma and palpitation of the heart; sick-headache; hysterics; apoplexy, etc. These maladies are so various, continually changing and complicated, that they require the constant watchful care of a family physician. If at any time he merits this glorious name, it is in a case where he has to guide, safely, the mother of a family, like a faithful pilot, through the dangerous rapids of her critical period. As to general rules we would advise great moderation in eating and drinking, particularly not to indulge in articles of a stimulating nature, such as wines, liquors and spices; the sleeping apartment should be airy, well ventilated; violent emotions and fatiguing exercise should be avoided, also exposures to inclement weather, wet feet, etc.

As to medicines, *pulsatilla* and *lachesis* are, in this period, the principal remedies, particularly for the irregular menses; every four or six days alternately, one dose (six glob.) will be sufficient. If other diseases occur during this time, the advice of the physician must be sought.

ABNORMAL EROTIC SENTIMENT.

Nymphomania.—We would gladly desist to treat of the above disease in a work of this kind, on account of being obliged to speak about matters so delicate, if we did not consider it to be our especial duty to inform the reader about a disease, which must be recognized in its commencement to prevent its further and fearfully demoralizing progress. It consists in an excessive increase of the erotic sense. As this disease may even befall children of a tender age, from nine to twelve years, it becomes an important duty of mothers to make themselves acquainted with its nature. It manifests itself in the desire to have the erotic feeling excited, and this is accomplished by various methods of rubbing and masturbation, which clandestinely is practiced; in such cases a certain degree of modesty still has been retained, yet it does not rest there. Soon the patients become perfectly frantic in this respect and lose all finer feelings of modesty and decorum, bent only on the one object of gratifying their criminal passion without shame or hindrance. In the height of this awful phrenzy, charity applies the word mania, nymphomania, to this disease, which, in its beginning, frequently was caused either by neglect of parents in watching their children, or by passionate conduct on the part of the latter, producing crime and destruction.

At the time of puberty, and sometime after, practices leading to this disease occur most frequently, and parents ought to be particularly careful to watch, at that period, the conduct of their children.

The exciting causes of these abnormities of the awakening erotic sense are enumerated as follows by Dr. Columbat: "Thus in populous cities, idleness, effeminacy, or sedentary life, the constant contact of the two sexes, and the frequenting of places where everything inspires pleasure, prolonged watching, excessive dancing, frivolous occupations, and the study of the arts that give too great activity to the imagination, erotic reading, the pernicious establishment of an artificial puberty, the premature shock of the genital system, the concentration of the sentiments and thoughts on objects which keep the genital system in a state of continual excitement; finally, a number of vicious habits and excesses of all kinds, which react more particularly upon the sensibility of the womb, an organ to be considered in the female as the very centre towards which all the morbific actions seem principally to have a tendency."

In young children, the cause of this malady must be foreign to real sexual development, as this latter has not advanced far enough to become a morbific power. In such cases we have found that the presence of ascarides or pinworms, either in large quantities in the rectum, or sometimes even in the vagina, may give the first impulse to the undue excitement created by the continued rubbing on those parts, which the child at first in self-defence has to commence, by the continuation of which, however, unfortunately, the erotic sentiment becomes prematurely developed and habitually stimulated. In older persons, of course, excuses of this kind are inadmissable, except where real disorganizations or idiotic formation of the brain compels us to think otherwise.

The consequences of this dreadful disease are fearful in the highest degree; they need not be introduced here, as our only object was to inform the reader of the existence of such a disease, which, in its commencement, might be successfully intercepted in its growth by proper care and treatment.

If a child shows by its actions that it is troubled with an irritation in the private parts, it is the duty of the mother at once to examine carefully, and if worms are the cause, to have them removed.

The use of cold water by injections in the vagina and rectum, also the douche on the hypogastric region, are recommended. In older persons, the treatment must consist of a mental as well as a physical training. In most cases the unfortunates can be brought back again to life and happiness, by being treated kindly but firmly. *Camphor* in small doses, is perhaps the only remedy which has a sufficient bearing upon all cases of this kind, to be recommended in a work like this. We omit to mention others, as it would carry us beyond the legitimate sphere of our intention; this disease belongs emphatically to the forum of the physician, not the lay practitioner.

ABSENCE OF EROTIC FEELING.

Anaphrodisy.—This abnormal condition of the female system is, in itself, neither a crime nor does it beget disease; yet its presence is frequently disturbing the pleasure and happiness of those who anticipated both in the matrimonial union. If, unfortunately, the wife should have little or no passionate excitement, the effect must be unpleasant in many respects. As it is impossible to go here into any farther explanation of a condition which varies so much in cause and treatment, we must content ourselves with the advice to those who suffer under anaphrodisy, that if they can overcome the modest reluctance so natural to the female character, to ask the advice of a physician; their wishes, in most cases, will be realized by a rational treatment.

STERILITY.

The impossibility for a woman to conceive, is either absolute or relative. It is very difficult to point out the precise cause of sterility during the life-time of a person, as many of the causes, particularly those of absolute sterility, lie in the absence or malformation of internal organs, which cannot be examined. In this uncertain state of diagnosis, one rule has to govern us, that is, to consider a sterility so long, as a relative or curable one, as we are not positively convinced of its absolute nature.

Various remedies have been at all times applied to overcome this impediment of nature; the ingenuity of man has tried to fathom the depths of existence, in order to restore that harmony of natural processes, which is generally the cause of happiness and contentment in this world. But in no other department of science have we been rewarded with so few satisfactory results. Nature seems to have reserved this field for her own action; at least she frequently restores a faculty in later years, which in younger days she had denied to her offspring. Yet we would advise the reader not to neglect the counsel of a physician on this subject; because, sometimes a judicious treatment removes the barriers laying in the diseased condition of the female. Hundreds and thousands of women have been cured of this defect, though a positive promise to that effect could not have been given. There are instances recorded where women have become pregnant for the first time after having been married for twenty years. A general constitutional treatment with homœopathic remedies, is perhaps the best calculated to remove sterility, and such a treatment must be instituted by a physician, who is competent of ascertaining by careful investigation the probable cause of the difficulty. Cases depending on uterine congestion, are mostly curable, and frequently yield to a well-directed treatment.

2. PREGNANCY.

As a state of purely physiological development pregnancy does not, of itself, imply disease as a necessary consequence during its duration. Yet the many ailments which accompany it at the present day, being the legitimate result of the complicated, unnatural conditions and habits of society, create and sustain the idea which people generally entertain, that a pregnant woman can never be free from one or the other disease. It is our duty, therefore, to give the reader, if not an extended treatise, at least a synopsis of those ailments commonly experienced during gestation, together with their treatment.

In the first part of this book, in the article on "Wife," we have stated all that we deemed necessary in regard to the physiological process of pregnancy, to which we must refer the reader, in order to understand many of the diseases peculiar to pregnant women.

PLETHORA—CONGESTION—FEVER.

These are forms of diseases to which pregnant women are more inclined than others, because in that state the blood has naturally a higher degree of plasticity or formative energy, which explains the fact that women otherwise weakly and badly nourished, frequently suffer in this direction; how much more must it be the case where women allow themselves to indulge in too nourishing and highly seasoned food or stimulating drinks.

Plethora manifests itself by a full, hard and more frequent pulse, rush of blood to the head, vertigo, buzzing before the ears, numbness of the extremities, asthma, sleeplessness, etc. In such a state it would be dangerous to await her confinement; various diseases might result from this plethoric condition, of which we will mention only one—convulsions—sufficiently important to induce energetic action against plethora during gestation.

Congestion and fever do not differ materially from the former in symptoms or their intensity; they appear upon the whole, however, more locally and are more transient.

The principal treatment, particularly to prevent plethora, consists in a rigid and careful regimen as to diet, exercise and habit. This, in most cases, will be sufficient to prevent its appearance, and obviate all further medication. For this purpose the patient must avoid all mental and corporeal excitement; eat little but often, more vegetables than meat; drink nothing but water or lemonades, if she is not inclined to diarrhea; dress not too warm and exercise freely in the open air, avoiding, however, fatigue.

If notwithstanding these dietetic rules, symptoms of plethora develop themselves, the patient should take, from time to time, *aconite* and *bella-*

donna alternately, every six, twelve or twenty-four hours a dose (six glob.) which, better than bleeding, will counteract the above named plastic quality of the blood, the real cause of plethora. It is not a surplus of blood which forms the characteristic of plethora; such a thing cannot exist in the vessels limited to a certain quantity; this surplus, if it existed, would sooner burst the vessels in the nose and lungs than be confined in too small a room. The above idea has only obtained on account of the means which were used to counteract the evil. Bleeding was heretofore considered the only remedy for plethora, and as it certainly gives momentary relief, the physicians judged the disease by the remedy, or rather by the action which the remedy produced. Subsequent investigations have shown that the benefit of a venæsection does not consist so much in the lesser quantity of blood remaining after it in the system, but in the nervous influence, and its reflection on the quality of the remaining blood. Bleeding constitutes, therefore, a remedy for this form of disease, although its application is limited to but few instances, particularly where plethora threatens to give rise to that formidable disease called puerperal convulsions, one form of which is frequently occasioned by a state resembling plethora, where an early venæsection will be beneficial, if the head is very much congested at the time.

HEMORRHAGES.

In the article on menstruation, we treated of the hemorrhages from the womb, which, if they occur during pregnancy, require a similar treatment, except in cases where symptoms of a miscarriage appear; for these the reader will find the necessary advice in the article on that subject. Here we intend to speak about the hemorrhages from other parts of the system, from the *nose*, the *lungs* and the *stomach*. They are frequently caused by a congestive tendency and have a critical import as to their cause. If they are not too severe at the time, the local congestion generally is diminished by them, and the patients feel relieved afterwards. But if they are too profuse and repeat too often, it is necessary to interfere.

Aconite should be used first, particularly if a feverish state preceded the hemorrhage; if it does not speedily relieve, alternate it with *bryonia*, in solution, every half hour a tea-spoonful. If there is however, not much fever, but more coldness, or if the bleeding has already continued for some time, give *ipecac.* and *china* in alternation, also in solution, in the same manner until better.

In slight cases the above prescription will be suitable for hemorrhage from the nose, stomach and lungs, in severe cases of bleeding from the lungs, the alternate use of *opium* and *ipecac.* or *china* will be necessary; in that of the stomach, *china* and *veratrum* or *arsenic*, and the application of cold water or ice on the root of the nose if the bleeding from this part does not yield speedily In regard to the vomiting of blood, we would yet remark that it constitutes, frequently, a symptom of inflammation of the spleen, and if so, the treatment of this disease must be instituted before the vomiting can stop.

HEMORRHOIDS—PILES.

This disease of the rectum frequently occurs during the latter part of gestation, and occasions, beside great annoyance and distress, sometimes even symptoms of threatening miscarriage. By carefully avoiding constipation, which may be considered one of the principal causes of this complaint, it can be mostly prevented or its attacks rendered mild and comparatively short. We can also by medicines given internally, and water, either warm or cold, applied externally, mitigate the severe pains and swellings, though we may not be able to cure this disease while gestation continues, which being its cause, will constantly re-produce it.

Nux vomica and *sulphur* in alternation, every evening a dose, (six glob.) are the principal remedies against it. If these should not relieve within a few days, recourse may be had to *ignatia*, if the pains, like violent stitches, shoot upwards and much blood is discharged or the rectum protrudes at each evacuation; it also quiets the nervous system, if irritated by the ineffectual straining to evacuate, so often experienced after a discharge in persons troubled with piles.

If constipation is present in a high degree, alternate *ignatia* with *opium*, every two or three hours a dose, (six glob.) until better.

If these remedies do not relieve, give the following. *Sepia, bellad., hepar, lachesis, arsenicum, carb. veget.* in their order, in intervals of a day; of each remedy two doses, (six glob. each) until better. Externally the application of cold water in sitz-baths, compresses or injections, is of the greatest benefit when the tumors do not bleed, but are very much enlarged and painful; while the warm water or steam is preferable when the tumors bleed or have ceased to bleed, yet continue to be very painful. Almost entire abstinence from food, except bread and water for a few days, shall, according to some authors, be an excellent dietetic in piles. Meat diet is to be avoided as much as possible. The removal of the tumors by the knife is a painful, and during pregnancy, highly dangerous operation, as thereby abortion may be excited. We, therefore, warn the reader never to have an operation of this kind performed during pregnancy.

VARICOSE VEINS.

Like the piles, this disease is caused by the pressure of the gravid uterus upon the lower blood-vessels, if it occurs during pregnancy; some females have a tendency to it all the time from a constitutional debility in the large abdominal veins. Such persons during pregnancy suffer severely from varicose veins, which sometimes enlarge to such an extent as to burst and discharge the blood either externally or into the cellular tissue. In such cases a horizontal position on a couch or bed, and bathing the limbs with an *arnica* solution, taking internally *arnica* and *pulsatilla*, once or twice a day, will be sufficient. The laced stocking or a bandage wound around the limb tightly from the foot upward, will be an excellent preventive for varicose veins. Beside the repeated application of cold water, bathing or washing is strongly recommended, as also the frequent rest on a couch or bed during the day, after having loosened the clothes around the waist.

The following medicines, if taken in long intervals, have a very beneficial effect to mitigate the disease. *Nux vom.*, *arsenicum*, *lachesis*, *lycopodium*, *sulphur*, two doses of each remedy within two weeks, each week one, until one after the other has been taken or the patient feels better.

SWELLING OF THE FEET AND LOWER LIMBS.

The same cause, which during gestation, produces varicose veins and piles, can also produce a swelling of the feet, the limbs and even of the whole body, as in some extraordinary cases it has been witnessed. By the pressure of the extended womb on the larger lymphatics, a stagnation ensues, which prevents the absorption of the lymphatic fluid, and causes it to accumulate within the areolar tissue. In the evening the lower limbs are generally swollen more, as the water sinks by its own gravity; in the morning the face and eye-lids show more of the watery infiltration. As long as no fever, at least no full, hard pulse attends a circumstance of this kind, it is of no great significance, save the annoyance and trouble to the patient. It disappears quickly after delivery, sometimes in a few hours; neither does it interfere to such a degree during delivery, as might be believed from the extent of the swelling. An immense absorption must take place, even during parturition, because in one case, which came under our own observation, no hindrance to a successful termination of the birth was experienced, although prior to its commencement, it seemed almost impossible that it could take place at all, in such a degree were the external parts swollen, almost closed. In that case, the absorption took place during the labor, as soon as by change of position, the child exerted no more pressure on the lymphatics.

But if the pulse is hard and full, measures must be taken to relieve the patient. She must take exercise, keep the bowels open by means of cold water injections, and the internal use of *bryonia* and *opium*, every two hours a dose, (six glob.) until relieved; if the fever rises still higher, take *aconite* and *bryonia* in the same manner.

TOOTH-ACHE.

This is a frequent complaint during pregnancy; it requires the same attendance as when it occurs at other times, with this exception, that the extraction of carious teeth should not be permitted, as the shock occasioned thereby might bring on an abortion. *Chamomile, belladonna, mercury, sulphur*, every hour a dose, (four glob.) until better, generally relieves the severest pains, and makes extraction quite unnecessary.

SALIVATION.

We had occasion, in the first part of this book, to mention salivation as one of the signs of pregnancy, at least in some women. Sometimes it continues during the greater part of gestation, and becomes one of the most troublesome and weakening ailments, particularly if it is combined with nausea and vomiting, when the derangement it creates, in the health of the woman, becomes truly alarming. While the future mother thus sometimes is brought to the verge of the grave, her expected offspring does not suffer in like manner, on the contrary, in most cases, appears to have done the better for it. This, by way of consolation to the sick woman, for whose comfort we cannot do much. Yet some of our medicines, even in this constitutional disorder, mitigate frequently to a great extent, the severity of the salivation. *Mercury* is one of the principal remedies, if salivation was not caused by the abuse of mercurial preparation; also *lobelia, lachesis, iodine, hepar, pulsatilla, sulphur, nitric acid*. Take each of the foregoing remedies on four consecutive evenings once, (four glob. at a dose,) discontinue a few evenings, to await its results, and if not better, take the next remedy in the same manner.

DERANGEMENT OF APPETITE.

This may manifest itself in various ways. Sometimes a complete disgust for every kind of food occurs, while in other cases, the appetite is so excessive as to become really a symptom of a diseased state of the stomach; again, in some cases it becomes capricious, desiring the strangest and most unusual articles for food, such as chalk, charcoal, etc. As it is almost impossible to correct this abnormal condition by the application of medicine, we must recommend to those suffering under it, to consult their own feelings as long as it is prudent and reasonable. Any excess must be detrimental.

To remain without food for too long a time, merely because we have an aversion against it, would finally lead to an incurable state of exhaustion. One article, ice-cream, a patient of this kind certainly can bear, if nothing else will stay on the stomach; she can at least sustain life with it. In some cases I have seen that Scotch herring, ale, champaigne, or other spicy articles, did overcome the sickly repugnance of the stomach. Anything is good which effects our purpose, and the patient must never tire to try, until she has found what will suit her case. *Arsenicum*, every other night a dose, (four glob.) will frequently restore appetite.

Those who indulge in the eating of strange and unusual articles, ought to be reminded that though a little to satisfy their craving might not do them any harm, a great quantity continued to be taken for a long time, will have a deleterious effect. Dr. Dewees relates a fact of this kind, where a lady died from the effects of eating chalk in too large quantities.

If the appetite becomes too excessive, the bad consequences for the system are not so trifling as persons might believe at first. The various symptoms of repletion, congestions to the head, lungs, and bowels, can take place; also head-ache, bleeding of the nose, lungs, etc., beside a disordered state of the digestive organs, not easily cured. In such cases, it is well to use food containing less nutriment in a greater bulk, such as rice, arrowroot, farina, etc., which is still very digestible. A little moral effort to restrain the excess of the appetite, is also very desirable, and ought to be practiced by reducing the quantity, not by abstaining from food entirely for some time. *Calcarea carb.*, every three or four nights a dose, (four glob.) will frequently curb the appetite within its proper limits.

NAUSEA AND VOMITING.

These are perhaps the most frequent and annoying ailments during pregnancy, and resist in most cases, the best directed efforts; medication can very seldom entirely subdue them, though an amelioration may be effected. As the nausea generally is greatest when the patient rises in the morning, the disease has received the name "*morning sickness.*" Its cause is as yet enveloped in the general mystery which hangs over the secret proceedings of gestation, and its sympathy with other functions. An increased uterine sensibility, reflecting on the ganglion nervous system, seems to be the excitor of a great many of these sufferings during pregnancy, and a proper, harder mode of life the best calculated to prevent them. However this hard, active life, in which manual labor forms the principal occupation, must not be commenced during pregnancy, but prior to it, in order to prepare the system for making the change within beforehand. This disease commonly ceases at the third or fourth month, very seldom continuing through the whole term. The nausea is frequently succeeded by the vomiting of a frothy mucus, sometimes even of bile, after it has continued for some time quite violently.

Various means have been tried to subdue this affection, but what in one case is successful, often fails in another. Generally we have found that riding in a carriage, even over a rough road, produces more comfort to the patient than anything else; fresh air and a voyage certainly have relieved some of the most distressing and obstinate cases. Beside this, select from the following remedies, they will mitigate at least the severest suffering.

We consider *tabacum* to be the principal remedy, particularly if, with nausea, there is fainting and a deadly paleness of the face, relieved by being in the open air; the patient loses flesh very fast; vomiting of water or acid liquid and mucus.

Sepia, if the nausea has lasted a great while, and appears mostly in the morning; in the evening there is a painful feeling of emptiness in the stomach, with burning and stinging in the pit.

Veratrum, very suitable after *tabacum*, or with it in alternation, particularly when the nausea is combined with great thirst; yet the patient vomits, after drinking ever so little, and has sour eructations, with great debility. If the latter symptom is present, *arsenicum* may be very beneficially administered alternately with *veratrum*.

Cuprum, if cramps in the stomach or chest appear during the paroxysm of vomiting, in alternation with *ipecac.*, if there is a tendency to diarrhea at the same time.

Camphor, in small doses, will frequently relieve, when a cold perspiration covers the skin, with hot head and cold extremities.

Secale after *veratrum* or *cuprum*, if pains in the uterus manifest themselves, similar to false labor pains threatening abortus.

The above remedies may be given in solution, every half hour or hour, a teaspoonful, until the paroxysm is over; afterwards only every four or six hours a dose, as circumstances require.

DIARRHEA.

This disease does not occur as often during pregnancy as constipation, but it is more injurious, because it weakens the system more, and needs, consequently immediate attention.

There is one form of it which needs an especial description. Women of a scrofulous constitution frequently are very constipated during the whole period of pregnancy, until a few weeks before confinement, when all at once a diarrhea commences, which lasts from eight to ten days. During labor, and a few days after, they seem to be entirely free from it, but very soon afterwards are attacked again, at which time the discharges, assume a purulent character, disclosing at once the presence of a fearful disease, ulceration of the bowels. At the same time, the secretion of milk has not been diminished, on the contrary, is increasing in quantity, and its quality rendered richer, so that the child thrives well, while the mother runs down, and if not relieved, will soon waste away under hectic fever. The mouth becomes sore, very tender, easily to bleed; at that stage it is called "*nursing sore mouth.*" This disease needs our closest attention. The strictest rest in a horizontal position, the mildest nourishment, such as farina, etc.; the exhibition of *nux vom.* and *hepar*, alternately in repeated doses, (every three hours,) generally relieves the patient in a few days, after which a few doses of *nitric acid* will be sufficient to finish the cure.

Other forms of diarrhea require remedies similar to those usually given, of which the reader will find the necessary information in my Domestic Physician, under the head of diarrhea. We mention here only the names of some of the most prominent. *Antimon. crud., phosphorus, pulsatilla, mercury, sepia.*

CONSTIPATION.

A sluggish condition of the bowels is a frequent concomitant of pregnancy, caused by a combination of circumstances, occurring at that period. A diet regulated especially to counteract it, the eating of cooked, or ripe sweet fruit of different kinds, such as prunes, apples, etc., and the drinking freely of cold water, besides active exercise in the open air, will be mostly instrumental to remove this difficulty. If these means should not succeed, recourse should be had to the following medicines, for the use of which a more specific reference can be found in my Domestic Physician, in the article on constipation. One of the best means to overcome long-continued constipation, is the alternate use of *nux vom., opium* and *platina*, every three or four hours a dose, (six glob.) until an evacuation is effected; if it should be too tardy or insufficient, an injection of cold water will aid the effect of the medicine. *Bryonia* and *ignatia* will frequently suffice to remove constipation, if the bowels feel painful, and *lycopodium* or *sulphur*, if it has continued for a long time.

DYSPEPSIA, HEART-BURN, ACID STOMACH.

These complaints, though not confined to pregnancy, nevertheless occur very frequently at that period, and particularly in those persons who were troubled with them previously. This fact is explained easily by the pressure which the enlarged uterus exerts on the stomach, especially in the latter months, thus interfering with digestion. *Nux vom.* and *pulsatilla* are the principal remedies to regulate these disorders; the former, if they are attended with constipation, the latter, if with diarrhea. For an acid stomach, frequent but small quantities of lemonade, or a mixture of one drop of sulphuric acid in a tumbler of water, is highly recommended, and will, sooner and more lastingly, correct the acid in the stomach, than lime-water or magnesia, which at best only absorbs the acid present, not preventing thereby its new formation.

DIFFICULTY OF SWALLOWING.

This may occur in any female of a nervous disposition, and at any time, not constituting, therefore, an ailment peculiar to this period; yet if it does appear during pregnancy, it becomes very annoying; though not dangerous, it sometimes requires our action.

The principal remedies are *belladonna, lachesis,*

and *ignatia* in alternation, every evening a dose, (six glob) until better.

SPASMODIC PAINS AND CRAMPS,

In the Legs, Back and Stomach. (Colic Pains.)

Pains of this kind are generally of a neuralgic nature, and occur mostly during the first half of gestation, though cramps in the legs are witnessed during the whole term. Their treatment does not essentially vary from that followed when present at any other time. In using external auxiliaries, such as warming bottles, hot bricks, blisters, etc., we would warn the reader not to apply them too hot, on account of the peculiar state of the patient, and if possible to do without them altogether, or to substitute bandages dipped in warm water, or a warm hip bath. Rubbing affords great relief, when the calves of the leg are cramped, or the pain in the back is very severe. The following medicines should be used, one at a time, dissolved in water, every half hour a tea-spoonful, until better; if this is not the case within one or two hours, the next should be taken in the same manner.

Cramps in the abdomen: *colocynth., nux vom., chamomile, ignatia, pulsatilla, belladonna, lachesis, veratrum, cuprum.*

In the legs: *veratrum, secale, cuprum, chamomile, sulphur.*

In the hips: *colocynth., rhus, belladonna, china, ferrum.*

In the feet: *calcarea carb., sulphur, graphites.*

In the back: *ignat., rhus, kali carb., bryonia, belladonna.*

DYSURY, STRANGURY, ISCHURY.

Scanty and painful urination are of frequent occurrence during pregnancy, caused not merely by rheumatic and gouty disorders, external injuries, suppressed piles, etc., as is commonly the case, but also by the pressure of the extended uterus upon the neck of the bladder, which makes urination difficult and painful. If this pressure continues too long or becomes excessive, the discharge of urine may be prevented thereby entirely, so that a complete *ischury* or *anury* takes place, which will require the application of an instrument, called catheter. Before the evil, however, grows to that height, the patient can try different positions, particularly in reclining, in order to relieve the neck of the bladder from the mechanical pressure of the womb.

If cold or rheumatism was the cause of the strangury, the patient will feel great relief from drinking freely of cold water, or slimy drinks, such as slippery elm, gum arabic, etc., which dilute the urine, rendering it less irritating to the bladder, and urethra. Besides, the following remedies should be taken in their order, every two or three hours a dose, (six glob.) until better.

For *strangury*: *nux vomica, pulsatilla, cocculus, cantharides.*

For *ischury*: *aconite, belladonna, cantharides, hyoscyamus, opium, lachesis.*

INCONTINENCE OF URINE.

A partial or total inability to retain the urine, is one of the most annoying complaints during pregnancy. Short and frequent hip baths, and cold water bandages are of the greatest benefit; besides the following remedies should be used in their order, every other day a dose, (six glob.) until better. *Belladonna, causticum, hyoscyamus, conium, arsenicum, lachesis.* Also have reference to what is said on page 423 of my Homœopathic Domestic Physician, on this subject.

JAUNDICE. (*Icterus.*)

This disease, well known by the yellow color it imparts to all the white parts of the body, is not dangerous of itself, but becomes only so by neglect, when other serious disorders, hectic fever, dropsy, etc., may follow. In our diagnosis of jaundice we must not depend on the darker appearance of the skin alone; this assumes frequently a darkish, almost yellow color during pregnancy, without jaundice being present at all.

The distinguishing features, which always indicate it, are constipation with whitish, almost colorless feces, urine of an orange color, and a dry skin, with slight remitting or intermitting fever. We mention this disease here, because during pregnancy particularly towards its close, it sometimes occurs, caused partly by mechanical pressure of the highly extended uterus upon the biliary ducts, partly by the symyathetic influence, which gestation evidently shows even from its beginning on all the digestive organs, especially the liver. It is stated that jaundice more frequently occurs in winter than in summer, and oftener in blondes with a lymphatic, than in brunettes with a sanguine temperament.

Its cure consists in re-establishing, by degrees, the healthy action of the skin by means of an even, warm temperature in a room, or still better in the bed; frequent tepid sitting-baths, wet bandages around the stomach, and the sweating in the wet sheet are powerful auxiliaries in the treatment of this disease; constipation is relieved by cold injections.

Besides, the following remedies should be given; three times a day a dose (four glob.) of the remedy selected, for a few days until better. *Mercury* and *hepar* in alternation; *lachesis* and *sulphur* alternately; and if caused by a fit of passion *chamomile* and *nux vom.* in alternation.

PAIN IN THE RIGHT SIDE.

This pain, caused by a fullness or congestion in the liver, sometimes makes its appearance from the fifth to the eighth month, and is also the result of pressure and sympathy occasioned by the extended uterus on that organ. It mostly manifests itself as a deep-seated pain or aching, slightly increased by taking a long breath, and generally worse by laying on the right side, showing that the convex or upper part of the liver, next to the diaphram is the principal seat of the congestion. The patient also feels at certain times a marked sensation of heat, and of a dull, heavy weight in the part mostly affected. To be bled for this affection is not alone not beneficial, but really injurious, and is entirely discarded even by the old practitioners. The wet cold bandage, often renewed, gives more relief; besides a sufficiency of exercise, neither too much nor too little. The following medicines, however, will be beneficial to mitigate the suffering by dispelling the congestion.

Aconite and *mercury* in alternation every three hours a dose, (four glob.,) if the pain and heat are intense.

Chamomile in same manner, if the pains are of a dull, pressive character, with paroxysms of anguish.

Pulsatilla, if the pains appear like spasms and towards evening.

ASTHMA—CONGESTION OF THE LUNGS.

Mechanical as well as sympathetic causes produce, particularly in the latter months, congestions of the lungs, which affect the functions of the lungs and the heart. This occurs, however, more in such women who formerly have been troubled with similar complaints. Difficulty of breathing, asthma, palpitation of the heart, and sometimes a hacking cough with spitting of blood, are the immediate consequences of these congestions.

A great deal of care is needed on the part of the patient as regards diet and mode of living; what she eats must be of easy digestion, not much at a time, nor of a heating nature; she must avoid exposure to cold and dampness; attend carefully to the secretions of the skin, for which purpose she must dress warmer than common. If she is inclined to constipation, she should take frequent cold injections. Besides, give her the following remedies in their order, every three hours a dose (four glob.) until better.

Asthma.—*Ipecac. bryonia, belladonna, pulsatilla, arsenicum, veratrum.*

Palpitation of the Heart.—*Aconite, belladonna, pulsatilla, arsenicum, opium, veratrum.*

Spitting of Blood.—*Aconite, ipecac., opium, arsenicum, pulsatilla, arnica.*

Pleurisy.—*Aconite, bryonia, pulsatilla, arnica.*

Hacking Cough.—*Ipecac. ignatia, nux vom., capsicum, mercury, pulsatilla.*

VERTIGO.—CONGESTION OF THE HEAD.

Headache, fainting and vertigo are frequent and sometimes very distressing ailments during pregnancy. They are caused by the plethora and nervous irritability, which often accompany the functional processes during gestation; if they repeat too frequently and violently, they may even become dangerous. To prevent their recurrence, the patient must avoid all mental or physical excitement, follow the strictest hygienic rules in diet and exercise, and keep the bowels in a good state, either by eating relaxing articles, or using injections. If the feet are cold, while the head is hot, a cold foot bath every evening, with severe rubbing after it, is necessary; also a tepid sitz-bath in the morning, with rubbing after it. Almost the severest headache can be mitigated instantaneously by a hot hip-bath. Fainting is generally relieved by loosening the clothes around the waist, a draft of fresh air, and sprinkling the face with cold water.

Besides the above rules, the patient should take of the following remedies, every two or four hours a dose, (four glob.,) in their order, until better.

Vertigo.— *Aconite, belladonna, lachesis, opium, bryonia, nux vom., mercury, sulphur.*

Headache.—*Ipecac., belladonna, opium, bryonia, rhus tox., ignatia, pulsatilla, sepia.*

Fainting.—*Ignatia, chamomile, coffea, nux vom., pulsatilla, china.*

Sleeplessness.—*Coffea, belladonna, ignatia, nux vom., aconite.*

Depression of Spirits.—See Hysteria. Compare Homœopathic Domestic Physician, under the same headings.

NEURALGIC PAINS.

From the fact that the nervous system plays an important part in everything connected with gestation, it is evident that we can expect neuralgic disorders in the organical and functional sphere during that period. This manifests itself particularly in organs below the diaphragm, where the uterus exerts its greatest influence. Neuralgic pains in the abdomen are very frequent; they appear under the short ribs, near the hips, or in the region of the bladder; it is generally a dull, heavy ache or burning, stinging in the muscles of the abdomen, remitting or intermitting as regards intensity. If the pains are felt within the abdomen, they resemble colic pains. Sometimes the neuralgy extends to the muscles of the small of the back, of the lower limbs, of the neck, arms and head; and if it affects the diaphragm or the pit of the stomach, the patient is taken with fits of laughing or crying, with spasmodic eructations of wind, sneezing, yawning, stretching and sighing. All these different complaints proceed from the same cause, viz: excessive nervous sensibility.

If the extension of the uterus is very considerable, inclining too much in front, a circumstance which frequently causes distress, much may be done by applying a bandage around the abdomen for support. Besides, the patient should wash frequently in cold water, and rub diligently the parts affected with cold water, in order to strengthen them. For the different complaints under the above head, take the following medicines in their order, every two or three hours, a dose (four glob.) until better.

Neuralgic Pains.— *Coffea, belladonna, pulsatilla, chamomile, bryonia, rhus tox., arsen., veratrum, sulphur.*

Spasmodic Laughter.—*Belladonna, hyoscyam., phosphorus, stramonium.*

Spasmodic Crying — Sobbing. — *Pulsatilla, aurum, ignatia, platina.*

Spasmodic Sneezing—*Aconite, rhus tox., silicea.*

Spasmodic Yawning.—*Ignatia, rhus, platina, natrum mur., sulphur.*

As neuralgia is frequently the result of indigestion, persons subject to it, should be particularly careful in their diet, avoiding substances of a heating nature, spices, coffee, tea, etc.

PUERPERAL CONVULSIONS.

(*Eclampsia gravidarum.*)

This is perhaps the most dangerous disease to which pregnancy predisposes, and on this account its treatment should never be undertaken save by a skillful physician. We have introduced it here solely on the ground to make the reader acquainted with the symptoms, particularly in the beginning of the convulsions, where a prompt and energetic action sometimes will arrest their progress.

Dr. Meigs speaks as follows about this disease: "It is a malady that is not met with every day—but it is one so horrible in its appearance, so deadly in its tendency, so embarrassing sometimes in its treatment, that, although it is not met with every day, it is solicitously expected, and probably obviated in many a case, which, but for such careful supervision, would swell its proportions in the statistical tables." We call these convulsions puerperal, because they can only affect in such a manner women advanced in pregnancy, during labor, or in the first days of the lying-in.

The convulsion is sometimes preceded by symptoms which more or less prognosticate its appearance, though sometimes it may occur without the slightest premonitory symptom, like a flash of lightning, the literal meaning of the word eclampsia. Dr. Meigs considers the following to be the most important premonitory signs: "Women with short necks; those who are fat; those who possess considerable muscular strength; those whose tissues are firm, solid, hard and unyielding; those that are of a sanguine-nervous temperament; those who have swollen feet and hands, and such as upon waking in the morning complain of numbness in the hands and bloating of the features, those who are affected with a feeling of great weakness or with loss of sensation in one side of the face, or in one of the members; those who have suddenly lost their hearing; those who have vertigo, headache, flashing of light within the eyes, dimness of sight, double vision or half sight; those who have sudden loud noises in the ears, and such as feel as if a violent blow had been received upon the head; those, furthermore, who labor under intense anæmia, with distension of the blood-vessels and heart," and we might add, those who constitutionally cannot bear the slightest degree of pain without a severe shock to the nervous system—"all such are to be held liable, and closely observed and protected."

Puerperal convulsions are similar in appearance to other convulsions; spasms of the muscles in every part of the system, foaming at the mouth, spasmodic breathing, and above all, unconsciousness. There are two kinds, the nervous and sanguineous, as they are called; in the former the face remains pale until towards the last, when it assumes a bluish color; in the latter, the face appears from the first very much bloated and swollen, dark red; this kind bears bleeding from the arm or foot in the commencement, while in the former it would not be advisable to bleed. In both it is advisable to use mustard drafts on the calves of the leg allowing them to draw for at least half an hour; to put cold water or ice on the head and have the bowels moved by one or more injections of salt and water; at the same time the exhibition of remedies must not be neglected.

Opium in repeated doses of a lower potency, if the face is dark, red and bloated, the breathing very loud and snoring, the patient entirely unconscious.

Bellad. and *hyoscy.*, if the face is less congested, and there is a higher degree of unconsciousness.

Stramonium in alternation with *belladonna*, where the face changes from paleness to redness and *vice versa*, the eyes appear squinting and from time to time the patient laughs spasmodically.

Chamomile, ignatia, lachesis are also recommended.

MISCARRIAGE. (*Abortion*.)

Miscarriage can take place at any time between the first and seventh month. If the expulsion of the child occurs after the seventh and before the ninth month, it is called a premature birth, as at this period the child generally can be saved and the object of pregnancy be gained. The more advanced pregnancy is, the less is the danger which may result from a miscarriage. The oftener a woman has miscarried, the more her constitution inclines to new misfortunes of the same kind in future pregnancies. Miscarriages are more liable to occur again at the same period during gestation at which the former or last one happened; if that period is passed without accident, the danger to miscarry diminishes. Miscarriage, therefore, can become habitual in some females; it also can appear epidemically, as it were, in so far at least, as in certain seasons the uterine congestion in general increases, (menses appear more frequently and profusely, hemorrhages ensue spontaneously, etc.,) and in the same ratio the possibility and occurrence of abortion multiplies.

If a woman approaches the time when heretofore she had the misfortune to miscarry, she must be careful not to provoke a return by walking great distances, lifting, running up and down stairs, riding over a rough road, etc. These practices may excite at any time a miscarriage in females who never have had any predisposition to it, how much greater is the danger in those already predisposed. A weakening, luxurious mode of living, late hours, great mental excitement are causes of miscarriage, and must be strictly avoided. In fact any important irregularity in diet or mode of living, may produce abortion; its causes are numerous and frequently even form part of a woman's every day habit. For instance, tight lacing, dancing at evening parties and the excitement so studiously sought in our modern society. All these artificial means of producing life's fleeting pleasures should be disregarded by one who soon will feel the more lasting joys of a mother; for the sake of reaching that exalted happiness, she should cheerfully forego for a short time, the fashionable and doubtful amusements of so-called fashionable society.

Yet there are some characters so destitute of all maternal feelings and so infatuated with the ease and luxury of high life, that they are sadly agrieved as soon as they know they are enciente. They know what is before them; they will have to deny themselves in many things; wealth, of which they have so much, is unable to buy a release from nature in that respect. Knowing that they cannot escape the natural destiny of woman, they frequently conceive the unnatural thought to deprive nature of its legitimate right by producing abortion artificially, either in exposing themselves to violent exercise or by taking certain drugs, which, as they have been informed, will excite an early miscarriage. It cannot be denied that a low state of morals like this, does exist, even in the highest circles of society, where wealth is abundant to sustain the gifts of nature, and such barbarous conduct cannot be excused by ignorance or poverty. But seldom, we may say never, can they accomplish their criminal purposes without endangering, in the highest degree, their own lives and health. No medicine which is powerful enough to produce abortion, can do so without first poisoning the whole system by passing through the stomach and the blood vessels of the abdomen to the uterus; it will sooner destroy the life of these organs, than affect the purpose in view. If such desire is already criminal, its execution must be doubly so, because it adds the attempt at suicide to that of destroying the life of an unborn infant. It is not so easy to sever the bonds existing between mother and child, if both are well and the former has no peculiar predisposition for miscarriage. It is this latter only which allows the most trifling accident sometimes to produce an abortion, while in females who have no such susceptibility, the severest misfortunes leave the infant in the uterus untouched and unhurt.

The premonitory symptoms of a miscarriage are chilliness, followed by fever and bearing down sensation, which afterward increase to labor-pains; cutting, drawing, mostly in the loins and abdomen. A discharge of bright-red blood either immediately issues from the vagina or coagulated dark blood appears from time to time. Frequent repetitions of these efforts of nature are usually necessary to expel the fœtus, varying in duration from two hours to two days. As soon as the above symptoms appear, even in a slighter degree, the patient must lie down, keeping perfectly quiet, without moving. She should be placed upon a mattress or anything harder and cooler than a feather bed. The room should have plenty of fresh air; doors and windows should be open for that purpose. Her drink must be cool toast water, lemonade, ice water, etc., and in giving it to her, care must be taken that the patient does not exert herself in receiving it; her position on the couch or bed should be strictly horizontal; it is rather better if her feet are slightly elevated. No stimulating food or drink is allowed except when complete exhaustion from loss of blood supervenes, at which time small quantities of wine may be given. (see *Flooding*.) The covering of the patient must be thin but sufficient. Great care should be taken to prevent officiousness on the part of nurses and friends, who by conversation and over-kindness, disturb the rest of the patient. Under no circumstances dare a conduct like this be tolerated; complete rest being necessary for the patient, we must procure it for her, even at the risk of giving offence to those who fancy to have the right of showing kindness to the sufferer by doing her positive harm.

If the attack is brought on by mechanical

injury, a fall, blow, mis-step, walking, lifting, etc., take *arnica* (twelve glob.) dissolved in half a teacupful of water, every fifteen or twenty minutes a teaspoonful until better, or until after the lapse of one hour another remedy becomes necessary.

Secale should be given next in the same manner as *arnica*, particularly in females who have previously miscarried; in older ones, or in those who have a weak and exhausted constitution, especially when the discharge consists of dark, liquid blood, and the pains are but slight.

China in alternation with *secale* when the loss of blood is considerable, and weakness and exhaustion evidently increases; when buzzing in the ears, cloudiness of sight and loss of consciousness ensue and the patient faints when raising from the pillow.

Hyoscyamus if spasms or convulsions appear, with loss of consciousness and discharge of light red blood.

Crocus if dark, clotted blood is discharged, increased by the least exertion, with a sensation of fluttering or motion around the navel.

Ipecac. alternating with *secale*, if with flooding there is nausea, fainting, cramping.

Belladonna and *platina* in alternation either at once or after *ipecac.* has failed to relieve, particularly when the pains are in the loins and bowels, severe bearing down, as if the intestines would be forced out; sensation in the back as if it were broken; face very pale or flushed; discharge of dark, thick, clotted blood.

After the fœtus has been expelled, the bleeding generally stops, and no further treatment is necessary, but the one commonly followed in a regular birth. The patient needs the same length of time to recover, and the same careful watching as during the lying-in. Indeed, it is advisable to keep in bed longer than common, in order to give the uterus a better chance to recover from the shock, and become stronger, also to avoid those diseases caused by weakness of the pelvic organs, such as prolapsus uteri, fluor albus, etc., which so often follow a miscarriage or regular birth if badly treated. Miscarriage must not be considered as a slight disorder; its consequences upon the system are sometimes fearful. As an interruption of the natural order, we can easily imagine that its effects upon the system must be more distressing than a regular birth, which is but a fulfillment of a natural development.

3. PARTURITION.

We have already, in the first part of this book, treated of delivery as a strictly physiological process, not requiring any more interference than is necessary for the sake of comfort and cleanliness. There are, however, many circumstances connected with this process, which formerly were considered and treated so differently from what they are at the present day, that it is very desirable to make the reader acquainted with the most approved methods and practices now in vogue during and after parturition. What we intend to say on this subject is not merely a recital of opinion, but has been found practically efficacious by us in hundreds of cases. Within the last thirty years the views on the treatment of women in labor and childbed have changed materially, and we are happy to say decidedly for the better. Before that time it was considered dangerous to allow fresh air or cold water to a woman in labor or after delivery, lest she contract diseases of all kinds. Now both are considered indispensable preventives and valuable remedial auxiliaries in the cure of those same diseases. Notwithstanding the slow progress which medical reform has made in certain classes of physicians, it is not to be doubted that in obstetrics even the most virulent opposers of reform in medicine have generally yielded with a praiseworthy zeal. Scarcely a physician of any denomination will be found, at present, advocating the old practices, whereby the natural process in labor or childbed is so much and so often interfered with, that disease and death may be the result

If a diarrhea precede the commencement of labor, as it sometimes does within the last few days of pregnancy, it is best not to interfere with it, as its effect will be rather beneficial than otherwise. It is usually caused by a nervous agitation, fear, etc., which frequently is relieved by looseness of the bowels. At the same time other nervous symptoms may manifest themselves, such as depression of spirits, a whining mood, a disposition to shed tears, fear of not surviving the labor and birth of the child, etc. *Aconite* and *opium* will be sufficient to quiet the nervous system; let the patient take, evening and morning, alternately a dose, (six glob.,) until better.

The real labor is near at hand, as soon as a slight discharge of mucus tinged with blood appears—the so-called " show"—which is a favorable sign at the beginning of labor.

The less we interfere with the action of nature during parturition, the easier will be its termination; the closer we follow the general laws of health during pregnancy, labor and childbed, the less sickness will attend women and infants. During labor our duty should be, in a great measure, directed to the arrangement of the outward circumstances, which exert such a great influence on its progress. For this purpose the bed of the patient should be placed so, that it allows of free access on all sides; it should consist, if possible, of a mattress, which is preferable to feathers. The room should be well ventilated, and provision made to have plenty of fresh air, when needed. The patient should have as much cold water as she desired; a denial in this respect would act very injuriously. We will now consider the different stages of parturition, their normal or abnormal condition, and treatment.

LABOR.

Labor is the term by which we express the process of nature to expel the contents of the womb. Labor-pains, or contractions of the womb, are the natural means to accomplish this object; they are consequently indispensable to the object in view, and a woman in labor should never consider her pains, although disagreeable, in any other than this light, else she might fret and whine herself sick without use, and thus deprive the most salutary provision in her condition of a part of its great benefit. Every expulsive effort of nature is connected with a certain degree of painful exertion, mingled with a feeling of ease and repose, which takes away, in a great measure, the severity of real pain. An analogous sensation may be experienced by the muscular contractions of the rectum and colon, which occur in intervals, thus permitting a grateful feeling of repose to intervene.

Labor may be divided into *natural* or *preternatural*, according to its own or the character of attending circumstances. If the labor is just sufficiently strong to expel the contents of the womb within a moderate time, without manual interference, and without danger either to the mother or child, we call it *natural* labor. On the contrary, if it is protracted, difficult, too painful or inefficient on account of its own weakness, or the presentation of parts which hinder its efficacy; if it is attended with other serious disorders, such as convulsions, hemorrhages, lacerations of the uterus or other contiguous parts, we call it *preternatural* labor.

Labor consists in the frequently repeated contractions of the uterus, aided by those of the surrounding muscles, by which expulsatory efforts the mouth of the womb gradually enlarges, the vagina widens and thus the child is forced through the extended passage.

From the above it is evident, that if no mechanical impediment either from the structure of the pelvis or the position or size of the child interferes, the labor will be sufficient in almost all cases to accomplish the object of nature without real danger. The time in which natural labor usually terminates varies; may be put down, however, from six to eight hours as the average duration.

The labor at the birth of a first child often lasts longer, but is on that account not more dangerous.

TOO SUDDEN TERMINATION OF LABOR.

Labor too quickly terminated, say in half an hour or less, is considered unfavorable for both the health of the mother and child. As regards the former, the chances for a sufficient and healthful contraction of the womb, so essential to avert danger of hemorrhage and after-diseases, are diminished by a labor too quickly terminated To prevent the bad consequences in this respect the woman should confine herself to a horizontal position for a longer time than usual, and apply frequent cool sponging-baths, in order to strengthen the parts, weakened by the too sudden termination of the birth The danger for the child consists in the mechanical injuries it may receive by falling on the floor, etc., if the labor should terminate so fast as not to allow the woman time of reaching a bed or couch. Cases of this kind have occurred; though very rare they may occur again. A woman having experienced too sudden labor, should the next time, towards the end of pregnancy, avoid exercising severely, and going up and down stairs; she should not omit to lie down frequently for a few hours and wash her stomach, back and hips once or twice a day in cold water.

By this process the muscular fibres of these parts near the uterine region are strengthened so as to offer a greater resistance during parturition.

PROTRACTED LABOR.

If the pains continue too long, say from twenty-four to thirty-six hours, labor becomes preternatural, and needs an especial treatment. We have to consider the causes which may produce this unusual delay. They are various, and most of them of such a nature as only an experienced midwife or physician is competent to remove. We will speak here of such as safely can come within reach of domestic help, being caused mostly by a mismanagement of labor in its early stages, or by a constitutional peculiarity of the patient, which can be remedied.

It is not unusual, however, that women in their first confinement experience a more than commonly protracted labor; this need not give rise to fear of danger, especially not if the woman is otherwise strong and healthy, as such ones more than the weakly, nervous women offer the greatest resistance to the expulsive efforts of labor.

If the woman has been urged to support the pains by exertions of her own at the very beginning of labor, thus exhausting her strength at a time when such efforts can be of no avail, it will frequently be found that towards the middle or close of the labor, when she ought to support it, she flags in strength, and from sheer exhaustion is unable to bear down or facilitate the expulsion of the child. In such cases, labor becomes protracted, and the only means of correcting it, consists in making the patient omit for some time her efforts until she has gained more strength; during this rest broth or wine in small quantities should be given to her.

If great restlessness has produced a high degree of nervous excitement, impeding often the progress of labor, the exhibition of *coffea* and *aconite* is indicated, in intervals of ten or fifteen minutes, a dose (four glob.;) *belladonna* should be given, if the head is hot and the feet are cold; and *nux vom.*, if the restlessness is connected with expressions of impatience. At the same time she should be exhorted to lie more quiet, and be covered well, in order to excite perspiration on the skin, when the latter is dry and husky.

If during the early part of labor, warm drinks (green, or chamomile, tea) had been freely given, by the use of which relaxation of the system and protracted labor had been produced, it is necessary to change, giving the patient cold water when thirsty, besides *coffea* and *china*, antidotes to green tea; *ignatia* and *pulsatilla*, antidotes to chamomile tea.

SUDDEN CESSATION OF LABOR-PAINS.

If they *cease* at once, either from fright or some other emotion, and the patient exhibits symptoms of congestion of the head, with red and bloated face, sopor, loud breathing, like snoring, *opium*, every ten minutes a dose (four globules,) should be administered. If upon the cessation of labor-pains, spasmodic distress in different parts of the body appears, *pulsatilla* should be given in the same manner.

If they cease in women naturally feeble and exhausted, *secale* is the best remedy to revive both the strength of the patient and the pains. It must be given in the same manner.

If protracted labor is caused by mechanical impediments, too narrow pelvic passage, abnormal position or unusual size of the child, manual assistance is necessary, which can be rendered only by competent persons. Convulsions occurring during labor very seldom retard its progress; they should, however, be treated forthwith, as stated on page 232.

SPURIOUS OR FALSE LABOR-PAINS.

We call such pains spurious or false, as do not exert an expelling power on the uterus, although resembling in other respects the true labor-pains, The difference between them is so great, that the patient herself cannot mistake it. The true labor-pains begin either in the lower part of the abdomen, in the region of the bladder, thence passing backward toward the spine, or they start in the back and shoot thence around to the front. Another still more certain sign of true pains is the cessation of all pain between two attacks, while false labor-pains never cease entirely.

Spurious pains, being caused by nervous or rheumatic disorders, can appear in any part of the system, but very seldom stay long in one place; they may be in the back, loins or front, wandering from one place to the other like rheumatism or neuralgy. They are soon dispelled by *pulsatilla*, every fifteen minutes a dose (six glob.,) the patient keeping as quiet as possible and well covered, to get warm and perspiring. Sometimes *chamomile* is necessary, if the patient does not perspire much.

Nux vom. is the most suitable remedy, when there is a pain as if from a bruise in the region of the bladder, or a constant but inefficient urging to stool; also when the woman is of an impatient, passionate temperament, having been inclined to the use of stimulants, wine, coffee, etc.

Bryonia, when the pains mostly lodge in the small of the back, and increase by motion, with irritability of temper, constipation, and congestion of the head; in the latter case alternately with *aconite*, particularly in women of a plethoric constitution, with a full, bounding pulse, flushed face, hot and dry skin.

Belladonna, alternately with *aconite*, particularly when the head is hot and the feet are cold; the pains resemble spasms.

Ignatia, when the pains are connected with great depression of spirits, and require frequent change of position, which mitigates their severity.

EXCESSIVELY PAINFUL LABOR.

The true labor-pains sometimes may become too violent, and following each other in quick succession, cause such great agitation and restlessness, as to render their mitigation necessary. This can be effected by the use of the following remedies.

Coffea and *aconite* alternately every ten minutes a dose (four glob.) until better, or the next remedy is indicated.

Chamomile, if the mind is greatly excited, the pains are intolerable; the woman very sensitive and impatient.

Belladonna, under the same symptoms, particularly when, with heat in the head, the feet are cold, is very beneficial in cases of first labor, where the unyielding state of the parts produces these extra exertions of nature.

Nux vom., in cases similar to *chamomile*, but with a constant, ineffectual urging to stool.

Towards the close of labor, the sulphuric ether may be applied to mitigate the severity of the pains; do not let a candle be near the ether.

THE WATERS. (*Child's Water.*)

At every birth a certain quantity of water is discharged during labor and after the child is born. This is the fluid which during pregnancy surrounds the child and is contained within the membranes, where it evidently has been placed by a wonderful provision of nature for the protection of the child as well as the mother. For the latter it is of the utmost service during labor, as it furnishes the best means of dilating the mouth of the womb in the beginning of parturition. When the mouth of the womb opens, the membranes are forced into the opening at every pain, forming a bag filled with the water, pressing on all points, evenly and gradually like an elastic wedge, possessing the mildest and surest power. As soon as the opening thus made, is large enough for the child's head to enter, the rupture of the membranes usually takes place, which causes so much of the water to escape as is contained in the bag, serving to lubricate the lower parts which now shall undergo the great distension. Much of the water, however, is still retained in the womb, which, during succeeding labor serves the same purpose, softening and lubricating the parts, and rendering the descending head moist and slippery.

Thus is this wonderful process, the birth of a child, facilitated immeasurably by the simple but effective agency of this limpid fluid. The immense benefit derived from the waters for this purpose, becomes sadly apparent when we have to witness labor which is deprived of its aid.

Sometimes the waters break too early and then they escape entirely, causing what is called a "dry birth." In such a case labor will always be very protracted and painful. To mitigate its severity, we have to supply what by accident or misfortune has been lost. We either inject sweet oil or thin, sweet goose-oil in the vagina, or apply it there from time to time on the points of the fingers.

This circumstance also teaches us the great lesson, never to break the waters too soon by artificial means. If it becomes necessary, we should wait until the opening of the *os uteri* is sufficiently enlarged to allow the head to enter when the water has escaped from the bag.

As the membranes around the child are composed of a double lining, it often occurs that between them a quantity of water accumulates, which is in no connection with the true waters contained within the inner membrane next to the child. This so-called *false* water frequently escapes long before the termination of gestation, in fact it can do so at any time, in which case it seldom fails to frighten the woman, exciting in her the apprehension of abortion or premature birth. The above explanation of this circumstance should dispel her fears, especially if she recollects that such a discharge of false water never is accompanied with labor-pains.

The quantity of the waters varies; it is sometimes very great, causing even suspicion of dropsy or the presence of twins.

DELIVERY.

The most agonizing so-called "cutting pains," occur just before delivery; they are of short duration, however, and of such a nature that, although extremely painful, they do not weaken, but, on the contrary, strengthen the patient, by arousing her energy to the utmost. At that time we frequently hear her cry out on the top of the voice, Oh, I must die! I must die! These words must not frighten those in attendance. It is true they are expressions of agony, but they do not indicate danger; they are rather of good import, signifying the effective progress of the birth. Kind, consoling words, stating that now the labor soon would terminate, etc., are the best remedy at this juncture of the case.

Just when the child is appearing, the attendant should support the region underneath its head to prevent a tearing of the tightly drawn skin at that point; this can be done with the open palm of the hand, pressed yieldingly underneath the head where it bulges out the most.

After delivery, but while the after-birth is still remaining within the womb, it is our duty to examine the uterine region outside, to make ourselves sure of the sufficient contraction of the womb, which, if rightly contracted, can at that time be felt hard as a stone, above the bladder, about the size of an infant's head. If not contracted, this lump cannot be felt, in which case we must either suspect another child within the womb or have to fear that by an internal hemorrhage, the womb has expanded again. In the former case the abdomen will feel quite large, and parts of the child yet unborn, are distinguishable through the walls of the abdomen, while in the latter, the womb will not extend as much but feel more yielding and soft. This last circumstance requires immediate, gentle, but firm and gradually increasing pressure with, and rubbing of the hand, which must be continued until the womb contracts anew, growing less in size and harder to the touch.

APPARENT DEATH, ASPHYXIA OF THE INFANT.

By this time the new born infant, still in connection with the after-birth, will have commenced crying quite lustily; if it should not have done so, it requires immediate attention. Its mouth must be cleaned from the mucus within, and its limbs and the navel string freed from all incumbrances. While one attendant procures warm water for a bath, another may improve the time by rubbing its spine, particularly that portion behind the upper part of the lungs from the neck downward, vigorously, and for a long time. This movement alone, is in most cases sufficient to restore animation and compel the infant to breathe. If no sign of life appears after fifteen minutes, put the infant, without cutting as yet the navel-string, into a warm bath, in which the rubbing on the spine and elsewhere should be continued. If this manipulation is unsuccessful, inflate the infant's lungs by breathing gently into its mouth, for which purpose it should be covered with a loose silk handkerchief, in order to break the force of the inflation. When the lungs are inflated, press the breast outside to expel the air again, and thus continue to inflate the lungs and expel the air in alternation for some time. If the child looks pale in the face, give it a dose (two glob.) of *tartar emetic*; if it looks bloated, almost crimson, give it *opium* (two glob.) on the tongue. If no signs of life appear after that, cut the navel-string if no pulsation any more is perceptible in it, (as long as the navel-string pulsates it should never be cut,) remove the child from the bath, dry off and wrap it well in warm flannels. After it has been allowed some rest, the efforts of re-animation should begin again in the same manner; in addition to the above, slight shocks of electricity now should be applied, directing the poles on the upper part of the spine and in front on the breast-bone.

AFTER-BIRTH.

If the child cries lustily, it may be removed soon by cutting the navel-string three inches from its body, but not before it has been ascertained that the navel string does not any more pulsate. As long as the beating of a pulse can be felt in any portion of the navel-string, it must not be cut; a few minutes detention on this account will never be of any injury, but may be of great benefit to the infant. The beating begins to cease first in that part of the string nearest the mother, and diminishes gradually towards the navel of the child. Wherever it has disappeared, the cord may be severed; even if it is too long at first it can be shortened afterwards at leisure. Before the cut is made, (generally about three inches from the navel of the child,) tie two strings made of firm yarn or cord, tightly around the navel-string, allowing an inch of room between them; in this place the cut should be made either with a pair of scissors or a good sharp knife, being careful, however, that no other parts of the infant's body are hurt by the operation.

After the child is removed, it may be ascertained how far the after-birth has been expelled, and whether far enough to allow of an easy removal. If, on examination, it is found to lie within the vagina, it can at once be grasped with the hand and extracted; but if the cord reaches higher than the upper part of the vagina, and its connection with the placenta cannot be felt, it is highly improper to seek its expulsion by force applied to the cord. The after-birth is expelled from the womb by labor-pains, and if these have ceased for a while after the expulsion of the child, without having effected that of the placenta, it is proper and more safe to await the return of the after-pains, which in due time will make their appearance. They may be hastened by rubbing and gently pressing the abdomen externally, over the region of the womb, exciting thereby this organ to new contraction. After a few pains of this kind, the placenta will often be found lying within the grasp of the fingers; at any rate but very slight tractions carefully made, will be sufficient to bring it down. This is all the manual assistance which persons not initiated in the art of obstetrics dare undertake as regards the removal of an after-birth. In most cases it should not be removed so quickly; there is at least sufficient time to wait for the arrival of a competent person. Meanwhile *pulsatilla* alternately with *secale*, every half hour a dose, (four glob.) may be given, which frequently will hasten its expulsion. If the patient's head is congested, face full and red, give of *belladonna* four globules, in preference to the above remedies.

HEMORRHAGE—FLOODING.

The only danger, perhaps, which may threaten women during and shortly after the birth of a child, is that which results from flooding. Yet, carefully managed, this will not occur very frequently, nor be so very dangerous. A great deal can be done to avoid it.

One of the most frequent causes of hemorrhage after delivery is mental excitement, either of a joyous or sad nature. Sometimes the husband, overjoyed at the safe delivery of his wife, expresses his own feelings in a too exciting manner, which causes like emotions in the fatigued patient; or the latter becomes suddenly very depressed in spirits on being told the sex of the infant, herself having expected it to be the opposite. Excitement of any kind must be carefully kept, at this period, from the patient. She needs complete rest of body and mind; sleep is the best restorer of strength, and the patient may indulge in it one hour after the birth of the child.

After delivery and the removal of the placenta, the woman should lie perfectly quiet for the first eight or ten hours, with the knees close together, well covered up to her chin, in order to keep up the perspiration excited on her skin during the preceding hard labor. A sudden cooling down sometimes produces a chill, with other bad consequences. When thus carefully covered, the nurse should wash the abdomen and other parts with moderately cool water without uncovering her, and if there is a tendency to a greater discharge from the womb than common, put a cold water bandage over the region of the womb, changing it whenever it becomes warm. Even if no flooding is threatening, a wet compress under the bandage commonly applied, will be a cooling and pleasant appliance, restorative in the highest degree, and preventing the abdomen from remaining afterwards too pendulous.

If *flooding* should occur, rub the region over the womb, using steady and powerful friction with the hand, until the womb contracts again, and after-pains appear, which diminish the danger of flooding; besides give the patient

Belladonna if she has a great deal of bearing down.

Chamomile if her limbs are cold and she has pains around the abdomen.

China and *ipecac.* alternately, in the worst cases, when the above remedies do not succeed; or

Pulsatilla, if a discharge of clotted blood appears at intervals, ceases and re-appears; followed by *crocus, platina* and *sabina*.

As the last and surest remedy, apply the coldest water in wet compresses, renewed every minute, or pounded ice on the abdomen; this will soon stay the flooding permanently, except when caused by parts of the after-birth being yet in the womb.

AFTER-PAINS.

Women, during the first confinement, experience, very seldom, after-pains. They are mostly caused by the renewed efforts of contraction in the womb, dilated from time to time by the clots of blood oozing from the parts where the after-birth adhered. This may last a few hours or many days, just as the case may be. The following remedies will mitigate their severity, particularly if they are of a rheumatic or spasmodic origin, as is sometimes the case.

Arnica is the first medicine, externally in a wash on the generative parts, and internally in globules; it sooths the irritability of the womb after severe labor. Alternately with it, give

Pulsatilla, every two or three hours a dose (four glob.) which regulates the uterine contractions.

Chamomile and *nux vom.* in alternation, in the same manner, if the pains are very pressing, producing impatience and irritability, with frequent but ineffectual urging to stool.

Coffea and *aconite*, in the same manner, alternately; if severity of the pains drive the patient almost to despair.

Secale in weakly females or those who have already had many children.

Belladonna, if the pains are attended with much bearing down, congestion to, and heat in the head, flushed face, coldness of the feet, tenderness and fullness of the abdomen, in alternation with *opium*, if the patient has an unusual disposition to sleep and stupor, during which her breathing is loud and snoring, only now and then interrupted by the severe after pains.

Sometimes the application of the cold, cool or warm shallow-bath, as circumstances may require, will be of great benefit; also the wet bandage around the abdomen.

CONFINEMENT.

This term signifies the lying-in of a woman for a certain length of time, during which she shall recruit and recover her former strength and health. In a perfectly natural state of society, and under circumstances where the physical power of woman is not marred, the time of confinement needs to be but very short. Indeed there are cases on record where women have been able the next day after delivery, to attend to their duties partially, and in a few days afterwards altogether. But such iron constitutions cannot be expected to exist in our artificial state of society. We are consequently compelled to adopt rules which will prevent the many disorders which may befall debilitated constitutions after delivery, if not carefully attended.

Our modern Hygiene has made great reforms, also, in this period of female development, and women are less longer now confined to their rooms than they were in former times, and we may, if we continue to live more in accordance with nature, arrive at still greater results. Water, fresh air and exercise will yet work miracles.

At the present time we hold it still necessary for the mother to remain in bed for the first five or six days, after which she may sit up awhile, at first in bed, gradually lengthening the time, until she can sit up at the end of two weeks altogether. Circumstances, of course, will modify the above. The greatest danger, however, which can arise in this period, is generally caused by excitement of the mind, variously induced. The principal fault in this respect, consists in allowing friends and neighbors to pay lengthy visits to the patient. Velpeau, a great obstetrician, speaks about this bad fashion as follows:

"It is important that the patient should neither speak nor be spoken to, except when necessary. A calm state of the mind and repose of the body are so indispensable, that too much care cannot be taken to remove every cause that might interfere with them. Most of the diseases which affect a woman in childbed, may be attributed to the thousands of visits of friends, neighbors or acquaintance, or the ceremony with which she is too often oppressed; she wishes to keep up the conversation; her mind becomes excited, the fruit of which is headache and agitation; the slightest and indiscrete word worries her; the slightest motives of joy agitate her in the extreme; the least opposition instantly makes her uneasy, and I can affirm that among the numerous cases of puerperal fever met with at the hospital de Perfectionnement, there are very few whose origin is unconnected with some moral commotion."

Before the fourteenth day, therefore, *visitors should not be admitted on any consideration. The risk is too great.*

The diet during this time, should receive our attention, but not with a view to cut short the allowance, or confine its quality to the mere dainties or so-called *light food*. Our opinion in this respect may differ somewhat from those who believe that a woman in confinement, although weakened by the whole process of delivery, by loss of blood and a great flow of milk, could subsist more comfortably, and gain strength, on light diet than strong, nourishing food. Our rule has been to let her desire alone in this respect, to let her choose her own diet, if no existing disorder will dictate otherwise. An healthy woman can relish and bear usual food the next day after delivery, as well as at any other time, and it is folly to make her starve at a time when she needs food the most It is well enough not to allow her to eat too much at a time, nor partake of any stimulating substances, either in food or drink; even tea or coffee may be better replaced by cold water. The room should be aired daily once or twice; the curtains around the bed should be such as to allow free circulation. Cleanliness in every particular should be observed.

The bowels are naturally constipated for the first four or five days after delivery. If, after the lapse of that time it is necessary to interfere, give a few doses of *opium* and *nux vom.*, in alternation every three hours a dose (six glob.); if head and limbs ache, give *bryonia* in the same manner. At the same time apply a few injections of cold water. Under no consideration whatever, give her cathartic medicine, the use of which is in no case more superfluous and hazardous. Stewed prunes or other relaxing dried fruit, will be equally efficacious and more harmless than physic.

LOCHIAL DISCHARGE.

After the delivery of the placenta, the womb does not immediately re-assume its former size and consistence; this reduction is the work of time, and, as it progresses, produces what is called the *lochia* or *lochial discharge*—liquids which ooze from the walls of the uterus into its cavity, whence they escape through the vagina. Through this process the womb gradually is reduced in size and its loose texture becomes firmer again, until a normal size and consistence is regained. The first two or three days this discharge consists of blood, partially coagulated; while the milk appears in the breasts, a more serous liquid is discharged, more or less tinged with blood; finally, on the sixth or seventh day, the flow becomes whitish or purulent, of a thicker consistence. This may last variously, from two or three to twenty days. It is evident, that a process like this can be disturbed either by general or local causes, operating upon the womb, from which, as from a sponge, these liquids are pressed. If the womb inflames or is congested, its contracting movement is impeded. consequently the discharge will cease; if the blood is constitutionally vitiated, its serum, as it oozes into the cavity of the womb, will be bad also, becoming sanious, thin, watery, of greenish color and bad smell, or a tedious suppurative process may take place, by which nature seeks to counteract the retarding influence of a bad constitution. Our remedies in such cases are intended to support struggling nature in the restoration of the natural discharge, as the best indication of the healthful action of the womb.

Irregularities of the lochial discharge during the presence of other diseases, such as childbed fever, etc., can only disappear after their cure, and require our attention in so far as they are symptoms of these diseases, to which we refer the reader.

Suppression of lochia, in consequence of congestion or inflammation of the womb, can occur after exposures to cold, errors of diet, mental emotions, sudden joy, fear or grief, etc.; chilliness, fever, sometimes delirium, thirst, headache, pains in the back and limbs generally accompany a suppression of the lochia. If high fever is present, with congestion of the head, delirium, etc., give *aconite* and *belladonna*, in alternation, particularly if there are delirium and violent pains in the head and back, with pressure in the genital organs, as if they would be thrust out, every two or three hours a dose (four glob.,) until better; if not relieved, give *bryonia* alternately with *aconite*, and if no improvement follows, the fever continuing, in alternation with *pulsatilla*, as above, especially when mental excitement of some kind or exposure to cold preceded the affection. *Veratrum*, after indigestion, with rush of blood to the head, delirium, and palpitation of the heart, every two hours a dose. *Opium* and *aconite* alternately, as above, if sudden fright was the cause. *Dulcamara* and *pulsatilla*, if no fever is present, and the suppression was caused by exposure to dampness and cold. *Coffea* and *chamomile*, alternately, if the patient is highly excited and unable to bear the pains; restless and impatient.

Warm compresses around the abdomen and a warm hip-bath are also recommended in cases of this kind. Diet the same as in fevers.

Excessive and protracted lochia generally require the same remedies as recommended for flooding. *Crocus*, if the discharge is dark colored, black, and of viscid consistency, with a feeling in the abdomen as of something alive. *China* and *ipecac.* in alternation, if the discharge appears in paroxysms, with nausea, vertigo, fainting, cold extremities, paleness of the face and hands, debility. *Calcarea*, in leuco-phlegmatic persons, fat, but flaccid, especially when there is an itching sensation in the uterus. *Belladonna* and *platina* in alternation, when the discharge is thick and dark, with drawing pains in the loins and abdomen; and *secale* in elderly and debilitated persons, with cool extremities and great anxiety of mind. *Rhus*, in cases where the lochia return after they once had ceased. *Silicea*, when the lochia appear each time that the infant is put to the breast.

The above remedies may be given as often as once or twice a day, until better, six globules as a dose.

Complete rest and good nourishment are indispensable to correct these disorders; the cure can be accelerated by shallow hip-bath of a medium temperature.

Offensive, sanious lochia need frequent washings with tepid water, and the use of *belladonna*, once a day a dose (six glob.;) if not better within eight days, give *carbo animalis* in the same manner, followed by *secale, china, carb, veg.*, if necessary, as above.

Diet must be very nourishing; patient must have plenty of fresh air, if possible, in high, dry locations.

CHILDBED FEVER.

There are few diseases more dangerous than this, particularly when it has been permitted by neglect or otherwise, to progress in its fearful career. Consequently, by introducing it here we cannot have the intention of enabling the reader to treat it, when fully developed; this stage requires the most skillful medical aid. Our aim is to make the reader acquainted with its character and symptoms, and such remedial means as will have a tendency to subdue it in the beginning, which can be effected more easily than its cure in after-stages.

Diagnosis.—Like most fevers, childbed fever is preceded by a chill, or at least a chilly sensation, crawling from the small of the back along the spine upwards; it is generally not very severe, even less so than those rigors indicating the approach of a milk fever, with which it might be confounded. But its apparent mildness and the presence of tenderness to the touch of the abdomen across the uterine region, which is wanting in milk fever, establish the diagnosis of approaching childbed fever. It is important to remember that tenderness of the abdomen is always connected with puerperal fever, even with the chill which precedes it. The slightest touch of the finger, nay, the weight of the bed-clothes on the abdomen is almost insupportable, the pains are sometimes so intense, that the patients cry out loudly and scream with agony. The mildest chill may be followed by the severest fever. The whole abdomen seems to be inflamed in a very short time; the lochia are suppressed, so is also the secretion of milk, if it already had appeared; if not, it will, of course, not make its appearance in that condition of the patient. Sometimes the childbed fever attacks shortly after delivery, within two or three days; at other times it appears only after ten or twelve days. The latter cases are considered more favorable than the former. If the pain is confined to one particular spot in the abdomen, the disease is not so dangerous as if the whole abdomen is painful to the touch. This latter is the most characteristic symptom of this fever; while other fevers may resemble it, having similar pains, none has this extreme tenderness to the touch, even to the slightest pressure on the abdomen. The pulse is always very frequent, as high as 150 per minute; excessive thirst, headache, burning fever, vomiting, etc., are present in most cases.

Without detaining the reader with a greater amount of detail on a disease, which, in its height, none but a professional man will allow himself to treat, we will indicate at once what has to be done in the beginning.

Treatment.—During the chill give the patient of *aconite* (twelve glob., dissolved in half a teacupful of water,) every half hour a teaspoonful, until it is succeeded by fever, when it is alternated with *belladonna*, prepared in a similar manner, every hour or two hours a teaspoonful, until better.

If the lochia have ceased, alternate *bryonia*, prepared similarly as *aconite*, until the severest symptoms cease. At the same time put around the abdomen a thick wet bandage, frequently renewed, until the heat disappears, and profuse perspiration, rest, quiet, and sleep ensue. Give as a drink plenty of cold water; fresh air is also necessary.

MILK-LEG. (*Phlegmasia Alba Dolens.*)

We make mention of this disease here, as one whose symptoms and course are so strange and frightful to the uninitiated, that a better acquaintance with its character is very desirable. Its treatment, when fully developed, must be directed by a skillful physician; only in the beginning, remedies may be at once applied to subdue it; these we intend to communicate.

Diagnosis.—The first symptoms of this disease are pain and swelling in some part of the leg or groin, which soon increase and prevent the motion of the limb at a very early stage. The fever present is generally not very high, but very constant. The disease makes its appearance usually within the first two weeks after delivery, and consists in an inflammation of the lymphatic vessels, veins and areolar tissue of the leg; the vulgar belief, that the milk, transferred from the breasts by some cause, appears in the legs, and makes them swell, is, of course, at the present day discarded as unfounded, and nothing is left of this belief but the popular name, *milk-leg.* It is astonishing to what an extent the leg can swell up in this disease, it being sometimes larger than a man's body.

Treatment.—*Belladonna* seems to be the remedy which, in the beginning, better than any other controls this disease. Give three times a day a dose (six glob.,) followed by *bryonia* and *rhus* in alternation, in the same manner, if necessary. As soon as possible procure medical aid; meanwhile keep the leg at rest and in an elevated position. The diet during this time must be very light, such as tea, toast and gruels; if a relish is wanted, lemonades, stewed prunes or other dried fruits are preferable.

MANIA IN CHILDBED.

The above name indicates the nature of the disease sufficiently; it is a mental derangement, caused by a peculiar condition in which the womb is during parturition and childbed. Although cases of this kind are rare, yet they occur; and if not known, might greatly frighten the attendants of the patient.

Sometimes the disease appears in the form of a *mania* with all the symptoms of rage, fury and wildness; frequently the head aches violently, the face is very red, eyes have a wild appearance, roll round and are very sensitive to the light; the pulse is very full and frequent, the patient generally very restless and agitated. In this state *belladonna* every two or three hours a dose (six glob.) will be beneficial, alternating it with *hyoscyamus*, if necessary.

At other times the disease assumes the character of a *melancholia*, without the exhibition of feverish symptoms; the patient appears low-spirited, talks at random, particularly when left alone; she is shy and given to fear, sleeps very little or none at all. *Belladonna*, *lachesis*, and *pulsatilla* are the principal remedies in this form of mania. They may be given, the first two in alternation, morning, noon and night, a dose (four glob.;) the latter twice a day a dose (six glob.,) if the former did not relieve.

As this disease, on account of its importance, requires the most skillful medical aid, we abstain from giving here more of its treatment. If the secretions, peculiar to childbed, such as milk and lochia, re-appear, a favorable issue of the disease may be expected.

4. NURSING.

With the termination of parturition the object of nature is only half fulfilled. The new being has been brought into the world; but torn from its parental roots, it would have gained nothing, if nature had not kindly provided the fountain of life, from which, for some time to come, it is destined to draw its nourishment, suitable in an eminent degree for its infantile nature. Without this breast of milk, the helpless young would perish by the hundreds; and though large, deplorably large, as at present is still the mortality of infants, it would be frightful, nay equal to a complete destruction of the human race, if nature had not mixed and prepared the food in the mother's bosom for the feeble offspring.

A provision of this kind demands our unqualified admiration and most loyal adhesion; although a law of nature in the fullest sense of the word, it does not manifest itself in passionate excitement for self-gratification, or vigorous egotism for self-preservation, which constitute the stimuli of most other laws of nature; but in the more divine garb of love, not in receiving but in dispensing blessings; not in defending one's self, but in protecting another one from starvation and death by offering for sustenance the life's fluid of one's self.

Sympathy and pity, therefore, the own Sisters of Mercy, who reside in heaven for evermore, move the tender heart of a mother, while nature swells her breasts with the sweet streams soon to be drawn in, with impatient delight by the most helpless being which nature produced, but which she also knew how best to protect and nurse.

For this purpose nature provided the milk-secreting organs whose function must be exercised, else injury will be done not only to the infants, by depriving them of their natural nourishment, but also to the health of the mother, by the distension and inflammation of the breasts. Reasons of the most urgent nature should only prevent a mother from suckling her infant. Serious diseases, such as consumption, eruptive or other fevers, great constitutional debility, mental derangement, etc., are some of those which demand a discontinuance of nursing.

But they occur very rarely; more frequently do we meet with a spirit of unwillingness on the part of mothers, to undergo the trouble, and occasional self-denial, which the nursing of infants forces upon them. In a majority of such cases, it is not want of maternal affection, but the belief that their own health and beauty might suffer, while no damage could result from it to the health of the infant by taking the milk of another one.

As to the first reason, mothers are sadly mistaken. All medical men agree that nursing, far from deteriorating or weakening the constitution, adds to the health and beauty of women. Besides, it is a matter of the greatest importance for the infant of having the most suitable nourishment; and none is more suitable than the own mother's milk. Neither in age or quality can it be rivaled by that of a wet-nurse, to say nothing of the many other disqualifying circumstances of the latter.

If it is, however, indispensable in some cases to have a wet-nurse, great care should be taken of ascertaining that she possesses the requisite qualifications. We would recommend that she should be examined thoroughly by the physician of the family, before she is permitted to give her breast to the child. She must be free from diseases of the skin, eyes and eyelids; she ought to have a clear complexion and healthy, full form. Her disposition should be mild and amiable; her character energetic, but not irritable. She must show a habit of cleanliness in personal matters; regularity and temperance in eating and drinking. Her morals should be above suspicion.

If these qualities have been found in a nurse,

it is necessary to compare the age of her milk with that which the child needs; it dare not vary three months either way. In all other respects, if she is engaged, she should live and act during nursing as the mother herself would do. It is well, however, to watch a nurse carefully for a month or two, before too much confidence is bestowed upon her, as she may slily do things not exactly wrong in her own opinion, yet objectionable altogether. Some nurses are in the habit of giving the children laudanum, to make them sleep well as they say. This, of course, has to be interdicted at once, for obvious reasons. Opium in any shape or form should be withheld from an infantile constition; it is more dangerous and destructive than alcohol.

If the mother concludes to nurse the infant herself, it should be placed on the breasts eight or ten hours after delivery, except urgent circumstances prevent it. In doing so, the nipples are at once drawn out and the act of suckling will encourage the flow of milk in the breasts, thereby preventing distension, as the milk already in the breasts being drawn out thus early, permits the newly secreted to take its place. Besides, the infant once having taken hold of the nipple while not yet made smaller by the distension of the breast, hardly ever refuses to do so afterwards, although the breast may be full and tense and the nipple almost disappear.

It is scarcely possible to state precisely the time when and how often an infant should take the breast; we can only advise the reader to be as regular in this respect as possible. Infants can be trained into certain habits very early. If no sickness prevents, an infant should have the breast about every three hours during the day time, while in the night it can do without the breast for six or eight hours. This habit once formed, will preserve its health better, on the known principle that regularity in eating and drinking does the same in adults.

It is a very reprehensible practice in mothers, to give the breast to children on the slightest occasion; for instance, when it awakes or begins to cry, either from being frightened, as children often are, or from real pain. An infant quieted by the breast, will soon go to sleep, during which a congestive state of the brain is produced, sometimes to such an extent as to engender spasms. It is much better to calm down the irritability of a child by more rational means; if nothing will help, a warm bath or some of the medicines recommended for that purpose in the Domestic Physician, will do it; *coffea, chamomile* and *belladonna* are the principal remedies for that purpose.

It is a question yet open for discussion, to determine when nursing shall cease. The opinions of medical writers are very much divided on this subject. Our own coincides with those who believe that it should not last longer than a year, certainly not over eighteen months. Others maintain that two and even three years is a time not too long. We hold that nature has indicated in the development of the child, her own wish as to the proper period of the termination of nursing. As soon as the teeth have appeared, the child is evidently ready to masticate and digest substances more solid than milk, and the eagerness with which it seeks to get hold of more solid food at that time, proves clearly that nature designed to terminate nursing.

It is proper, however, to continue it until the period of teething is fairly over, a time, during which the children are more or less delicate and feeble.

The diet of the mother, while nursing, should be more nourishing than common, although it is not necessary to be too particular in its selection. Her own wishes will generally point to the kind of food most wholesome for herself and child. As regards the latter, a little experience will soon teach the proper medium, and this must be kept. Acid food or drink, though perfectly agreeable to the mother, usually disagrees with the infant; consequently the mother will have to abstain from it. We have mostly found it the best plan to let nursing mothers prescribe their own diet, choosing it according to their liking; as to quantity we would advise them not to indulge their appetite too much at a time. To eat frequently but little at once, should be the rule. A so-called stimulating diet is under no circumstances advisable. Either the nursing mother is well, then she has no need of stimulating food or drink; or she is sick, then she needs medical treatment, which, according to homœopathic principles, is always without stimulus. We will now consider the various disorders belonging to the nursing period.

MILK FEVER.

Milk may appear in the breasts long before the termination of pregnancy; usually it makes its appearance the third day after delivery, in most cases gently and without much disturbance of the general health; but sometimes under a storm of excitement in the vascular and nervous system, which is called *milk fever*. After a severe chill, which penetrates the whole body, a violent fever appears, with headache, congestion to the brain and chest, which produces difficult respiration; during this time the breasts begin to swell and the patient perceives the *shooting in* of the milk; the breasts become now tender and hard.

Although this fever soon may disappear, it is desirable to mitigate its severity, which can be done by a few doses of *aconite*, every hour or two a dose (four glob.) As soon as the patient begins to perspire, most of the above symptoms will cease; if they should not, give *bryonia*, if the head and back still ache or there is great oppression at the chest. If the head, however, is the principal seat of distress, particularly if the patient cannot bear the light, give *belladonna* in the same manner. If the breasts are very tender to the touch, the patient is very restless and much excited, give *coffea* and *chamomile* in alternation, every hour a dose (four glob.,) until better. *Pulsatilla* is particularly indicated when the breasts are very much distended, feel very sore, and rheumatic pains extend to the muscles of the chest and shoulders; it should be given every three hours a dose (four glob.) Give *bryonia* and *rhus* alternately, every two hours a dose (four glob.,) if the tongue is coated and the back and limbs continue to ache, after the fever has disappeared.

External applications are of not much use during a milk fever, except, perhaps, a compress dipped in hot water and wrung out well. The milk should be drawn out as soon as possible, either by the child or a breast pump. Of the latter, we have found those having an India rubber globe attached to them, to be the best kind.

AGUE IN THE BREAST.—GATHERED BREAST.

It is a general law of our nature, that organs which are at certain periods especially active, are at that time more subject to diseases than others not equally active. During lactation, the breasts are the organs predominantly active in the female system; they are, therefore, the ones on which the diseases occurring during the whole of that period will reflect more or less. If a nursing woman gets cold, it will settle there; if by mental excitement, passion, fright, anger, fear, grief, etc., she becomes sick, it will affect first the secretion of milk; in short, any disorder during lactation has its bad effect on this process.

Ague in the breast is the most common form by which this effect manifests itself, and if not relieved at once, will end in the suppuration of the mammary gland.

A more or less severe chill is followed by fever, generally accompanied by lancinating or shooting pains in the breast, whose secretion is mostly arrested at that time. This increases the size of the breast and if the milk is not removed, will predispose to inflammation and suppuration. As a general rule we recommend to give at once

Chamomile and *bryonia* alternately, every hour a dose (four glob.,) for four hours; after which discontinue for four hours, and let the fever pass off by perspiration, without giving any more medicine; if the fever, however will not disappear, or if it returns, give

Aconite and *belladonna*, particularly when the

breasts are swollen, hard, and very tender; externally, apply hot brandy cloths. If lumps remain in the breasts, rub with sweet oil, or lay over the breast a plaster made of beeswax and sweet oil.

If a *gathering* of the breast cannot be avoided, abstain from applying the warm poultices as long as possible, as it has a tendency to implicate a still larger part of the breast within the suppurative sphere; give during this time

Phosphorus and *hepar* alternately, morning and evening a dose (four glob.,) until better, or until four doses of each are taken, after which discontinue the medicine, awaiting its effects at least three or four days; if no signs of improvement are visible, give

Mercury and *lachesis* in the same manner; and then again,

Phosphorus and *hepar*, until the abscess has opened or the swelling is diminished.

After the opening of the abscess and the discharge of the matter, give *silicea*, every evening and morning a dose (four glob.;) externally apply a wash on the breast three times a day, made of twelve globules of *silicea* in half a teacupful of water. This remedy may be followed in two weeks, if necessary, by *sulphur*, internally in the same manner as *silicea*, and thus in alternation with it, until the breasts are healed.

During this time poultices of bread and milk,

or slippery elm, may be used to mitigate the irritation. Let the diet be nourishing but not stimulating. The infant may be allowed to nurse as long as possible; and if not, use the breast-pump to draw the milk out as often as needed.

DETERIORATION OF MILK.

A good quality of human milk should exhibit a whitish color, with a tinge of bluish or yellowish; should taste pleasantly sweet and have no smell; a drop of it put on a nail should glide off from the same, if held in an oblique position, slowly, leaving a whitish mark on the nail. A drop of good human milk put in a tumbler of water, will mix in it slowly, forming clouds in it here and there.

Sometimes the milk deteriorates, becomes too thin and watery, or too thick, oleaginous, acrid, even acid; at other times it appears mixed with pus and blood, tastes bitter and assumes an abnormal color.

Milk can, however, be deteriorated without exhibiting any other external signs, save its bad consequences on the child. This is particularly the case after severe mental emotions, fear, grief, anger, etc, or the use of large doses of medicines, which, absorbed by the vessels, come into the circulation of the system. As to the effect of the mind on the secretion of milk, Carpenter remarks:

"The formation of this secretion is influenced by the nervous system to a greater degree, perhaps, than that of any other. The process may go on continuously, to a slight degree, during the whole period of lactation; but it is only in animals that have special reservoirs for that purpose, that any accumulation of the fluid can take place. In the human female, these reservoirs are so small as to hold but a trifling quantity of milk; and the greater part of the secretion is actually formed whilst the child is at the breast. The irritation of the nipple produced by the act of suction, and the mental emotion connected with it, concur to produce an increased flow of blood into the gland, which is known to nurses as the *draught;* and thus the secretion is for the time greatly augmented. The draught may be produced simply by the emotional state of the mind, as by the thought of the child when absent; and the irritation of the nipple may alone occasion it; but the two influences usually act simultaneously. The most remarkable examples of the influence of such stimuli on the mammary secretion, are those in which milk has been produced by girls and old women, and even by men, in quantity sufficient to support an infant. The application of the child to the nipple in order to tranquillize it, the irritation produced by its efforts at suction, and the strong desire to furnish milk, seem, in the first instance, to occasion an augmented nutrition of the gland, so that it becomes fit for the performance of its function; and then to produce in it that state of functional activity, the result of which is the production of milk. It is not only in this way that the mammary secretion is influenced by the condition of the mind; for it is particularly liable to be affected as to quality by the habitual state of the feelings, or even by their temporary excitement. Thus, a fretful temper not only lessens the quantity of milk, but makes it thin, serous, and gives it an irritating quality; and the same effect will be produced for a time by a fit of anger. Under the influence of grief or anxiety, the secretion is either checked altogether, or it is diminished in amount and deteriorated in quality. The secretion is usually checked altogether by terror; and under the influence of violent passion it may be so changed in its character, as to produce the most injurious, and even fatal consequences to the infant. So many instances are now on record in which children that have been suckled within a few minutes after the mothers have been in a state of violent rage or terror, have died suddenly in convulsive attacks, that the occurrence can scarcely be set down as a mere coincidence; and certain as we are of the deleterious effects of less severe emotions upon the properties of the milk, it does not seem unlikely that in these cases, the bland nutritious fluid should be converted into a poison of rapid and deadly operation."

There is evident danger in allowing the child to suck immediately after violent emotions of the mind; the same has been experienced after violent bodily exercise, running or performing hard manual labor. A suitable length of time should elapse before a mother dare give, with impunity, the breast to a child after such disturbing influences have occurred, and not even then should the child suck until after a considerable quantity of milk has been extracted artificially, else it will become sick from the milk which was in the breast during the excitement.

The quality of the milk can be improved by the use of the following remedies. If it is thin and serous take *china*, every other evening a dose (six glob.,) for eight days; if not improved at that time, take

Stannum in the same manner, and if not better after its use, take

Mercury, particularly if the infant refuses to suck.

If the color of the milk is too yellow and the taste a bitter one, give *rheum* in the same manner.

If the child throws up the milk *immediately* after sucking, give *silicea* every three or four days one globule, until better.

SUPPRESSED SECRETION OF MILK.

The circumstances which suppress the secretion of milk in the breasts, are as various as the manifold relations, external or internal, which influence our system. The most prominent, however, are the following: Exposures to cold or dampness, errors in diet, sudden and violent mental emotions, diseases in other parts of the system, particularly if they are of a nervous character. The consequences following a sudden suppression of lacteal secretion, are frequently of a serious nature, and their prevention requires our immediate attention. The danger is greatest when congestions to the head, breast or abdomen appear. Give immediately of

Pulsatilla (twelve globules dissolved in half a teacupful of water,) every two or three hours a teaspoonful; this will frequently restore the flow of milk, particularly if cold was the cause.

But should congestions of the head, lungs or abdomen be present, give

Belladonna and *bryonia* in the same manner as above, in alternation, until better.

If *mental emotions* have caused it and the patient is very much excited and restless, give first *aconite* and *coffea* in alternation, as above.

If exposure to *cold* or *dampness* produced the suppression give *chamomile, bryonia* and *rhus,* particularly when the head and limbs ache and fever is present, having been preceded by a chill.

If *diarrhea* sets in, give *pulsatilla, mercury, bryonia, rhus.*

EXCESSIVE SECRETION OF MILK.

A too copious secretion of milk may produce swelling and inflammation of the breasts with all its concomitants; also obstructed or involuntary emission of milk, debility, nervous and inflammatory disorders, headache, hysterics, even tubercular consumption. In such cases medical aid should be sought at once. Until that is procured, the following remedies may be given.

Belladonna every other evening a dose (six glob.,) will diminish the secretion, if febrile and congestive symptoms are present.

Calcarea carb., if the former does not give relief, in the same manner for one week. If not relieved, take

Phosphorus in the same manner; besides apply externally cotton batting, which mitigates swelling and pains.

The same remedies are beneficial, if the milk escapes all the time, keeping the parts constantly wet and rendering them more liable than usual to cold on the slightest exposure.

China should be given when great debility is present, either when the milk flows involuntarily or is secreted too copiously; in the former case it alternates well with *pulsatilla*, in the latter with *rhus*, every evening a dose (four glob.,) until better. Frequent washing and bathing is recommended.

DEFICIENCY OF MILK.

Various circumstances may cause a deficiency of milk in quantity; they are either constitutional or occasional. To the former belongs the mother's age, which if too young or too far advanced predisposes to this complaint; women inclined to corpulency have not much milk, also such as in former confinements, compelled by disease or otherwise, did not nurse their children; organic diseases of the breasts themselves may sometimes prevent the secretion of a sufficient quantity of milk. As exciting causes may be accounted depressing mental emotions, sudden change of the mother's mode of living, sedentary habit, exposure to cold and dampness, faults in the diet, abuse of cathartic medicines and a high degree of physical debility. As to the treatment of this disease we have, of course, first to remove its cause, as far, at least, as possible. After this is done the following remedies should be given to correct the remaining derangement.

Agnus castus is recommended very highly in cases of this kind; the patient takes every other evening a dose (six glob.) If constitutional causes operate, the patient should undergo a proper treatment by a competent homœopathic physician.

Calcarea, iodium, causticum, sepia and *sulphur* in such cases will be most beneficial.

Asafœtida in small quantities, either in tincture or first trituration, is recommended very highly; as also the *anis* and *dill seed;* the latter made into a tea and drank three or four times a day. Some women use drinks made of milk and various spirituous liquors; we would request the reader to be careful in their use, as the child will thus receive milk more or less impregnated with alcohol and its fearful consequences.

SORE NIPPLES.

This affection, consisting of an excoriation of the skin around and on the nipple, usually appears soon after the child has been put to the breast. Its main cause consists in a constitutional tenderness of the skin, which manifests itself on the slightest occasion, in cracks and wounds, of a very sensitive nature. We have witnesssed distressing cases of this kind; the sufferings seem sometimes to be insupportable, if the complaint has progressed very far. Our aim from the beginning should be to prevent the soreness of the nipples, by washing them and the breasts a few months before parturition twice a day in cold water. This strengthens the skin and renders it less liable to be affected afterwards. After the birth of the child follow the following directions. Apply

Arnica, six drops of the tincture to a teacupful of water; wash with it every time after the child has sucked. At the same time give internally

Chamomile, every four or six hours a dose (four glob.,) particularly when the nipples inflame, swell and threaten to ulcerate, with pains almost insupportable, like tooth-ache.

If this fails, give the following medicines in their order, each one once a day for six or eight days until relief is obtained.

Mercury, sulphur, silicea, graphites, lycopodium, calcarea carb. These remedies are intended to remove the constitutional taint which underlies the affection, and without the extinction of which no true cure of sore nipples can be effected.

One of the best expedients externally applied, is a cow's teat fastened on a silver plate; this instrument can be had in the drug stores. They are preserved in diluted alcohol and are washed in clean water each time before being applied. If the wound is not ulcerated, it may be covered with collodion, which permits the sucking of the child without tearing it open every time afresh.

CHAPTER II.

DISEASES OF GENERATIVE ORGANS.

The diseases to which the female organs of generation are liable, comprise some of the most difficult and dangerous disorders; they are very numerous and of frequent occurrence. It is very desirable, therefore, that the reader should acquire such general knowledge of their appearance and course, as will aid in preventing danger and destruction so often the natural consequence of neglect and ignorance. We intend to describe in the following pages, a number of those diseases which occur most frequently and whose progress can often be arrested in their commencement. Of most of them the reader must be content to learn only the description, as their treatment is either too difficult or too hazardous.

The organs of generation are divided into external and internal; we begin with the former.

IMPERFORATION OF THE HYMEN.

Sometimes the entrance into the vagina is closed entirely, which before the years of puberty is of but little consequence; it becomes important however, afterwards for very obvious reasons, because the menses are thereby prevented from making their appearance. As this difficulty is easier removed in infancy, it is well for mothers to observe early, whether it exists. We once had occasion to correct this fault in a child two years old; the operation was slight and without pain. If it is left until the age of puberty, severe pains are experienced by the girl every month, until, by an operation, a sufficient aperture is made to allow the menses to appear. When early recognized, the disease is not dangerous.

INFLAMMATION OF EXTERNAL PARTS.

From the structure, texture and position of these parts, it is evident that they must be liable to various and severe diseases, particularly if in addition to exposure, neglect of proper cleanliness acts as an exciting cause. Daily washings with cold water, and during menstrual flow with tepid water, are indispensably necessary to prevent these parts from becoming the seat of annoying and dangerous diseases The labia in young children sometimes are found adherent, which may be owing, frequently, to a want of cleanliness. An early application to a surgeon will correct this difficulty. Afterwards, however, these parts should be washed often to prevent further adhesions.

Inflammation of the labia occurs frequently, and has a' great tendency to terminate in suppuration, producing an abcess. The patient should, from the beginning, keep quiet in a horizontal position and take *aconite* and *bryonia* alternately, every six hours a dose (six glob.;) externally she should apply cold water compresses, if the swelling is not very large and hot; but if it is so, use hot water applications. If, notwithstanding, suppuration should ensue and the abscess form, *mercury* and *lachesis* in the same manner, are indicated. After the abscess has discharged, one or two doses of *hepar* will be beneficial.

WOUNDS ON THE EXTERNAL PARTS.

All wounds on these parts need more than common care; rest in a horizontal position is indispensable. Besides the application of *arnica* tincture in cold water, in compresses, kept wet constantly and laid over the parts affected.

OEDEMATOUS SWELLING OF THE LABIA.

Pregnant women are more liable to this complaint than others; sometimes it occurs in consequence of anasarca, when the lower extremities are already swelled to the greatest extent.

A swelling of these parts during pregnancy soon disappears after, or even during parturition; very seldom is the successful termination of the birth impeded by it. The patient can greatly relieve herself by lying down frequently during the day time for an hour or two. If the delivery is retarded by the swelling, the labia may be punctured on their external surface and the water let out, even after labor has commenced.

PRURITUS.

Itching of the Private Parts.—This is perhaps the most distressing and troublesome disease to which females are subject; it takes away rest and sleep, thus producing sometimes the most extreme debility. We are as yet unable to determine the precise cause of this disease; in most cases, however, it must be the secretion of some acrid fluid, which is discharged on these parts at intervals, when the itching takes place. Want of cleanliness may aggravate the attack, but scarcely ever can be its sole cause. Pruritus more frequently attacks pregnant women, but is not confined to them alone. At first its appearance may produce in the mind of the sufferer a suspicion of being affected with some secret disorder, thus adding mental sufferings to the intolerable physical ones, particularly if the disease appears in the form of aphthous eruption, like the thrush of infants.

In such cases a strong solution of *borax* in water, applied three or four times a day, if necessary by a syringe, will remove the itching in a short time.

Another very excellent remedy is the injection of *ammoniated* water into the vagina; it is best applied alternately with that of the *borax* solution.

The constant application of cold water is necessary in a few cases where the itching is more in external parts; it affords sometimes the only means to procure rest and sleep for the patient.

We should never omit to institute an internal treatment; the following remedies, to be given in their order, will be of great benefit; each remedy should be used for three days, twice a day a dose, (six glob.,) until better.

Apis mellific., *arsenicum*, *rhus*, *bryonia*, *pulsat.*, *mercury*, *sarsaparilla*, *sulphur*, *sepia*, *silicea*, *graphites*, *carb. veg.*

Before leaving this subject, we will draw the attention of the reader to a precautionary rule in the external treatment of this disease. If a patient, afflicted with pruritus, has schirrous tumors in the breast, the external application of the *borax* for the pruritus should be preceded by the internal use of the remedies above mentioned; in a similar case coming under our notice, the schirrus began to degenerate into an open cancer as soon as the pruritus had been hastily removed by the external use of *borax* alone; the lady in question died soon after.

The appropriate internal remedy needs no borax or other such

DISEASES OF THE VAGINA.

The natural malformations of the vagina are very rare, and consist chiefly, as Dr. Denman says, "of such an abbreviation and contraction as to render it unfit for the purposes for which it was designed. The curative indications are to relax the parts by the use of emolient applications, and to dilate them to their proper size by sponge or other tents, or, which is more effectual, by bougies gradually enlarged." In a case which came under our observation, the dilatation was effected by the last mentioned method; the woman afterwards became pregnant, and was safely delivered of a fine large child without any more than common exertions in labor.

A complaint of more frequent occurrence is the *prolapsus* or falling down of the vagina; it consists in the inversion and depression of the front or back vaginal wall to such an extent as to form a marked protrusion, sometimes outside the entrance. Its primary cause lies in a relaxed or weakened state of the mucous membrane, which, with other exciting circumstances, such as long continued leucorrheas and irritations, frequent pregnancies and hard labors, abuse of spirituous liquors, late hours, etc., permit the walls of the vagina to sink down and invert. It may be distinguished from prolapsus uteri, by the absence of hardness in the fallen or protruded lump, the uterus always exhibiting to the feeling a consistent, even hard surface, while the vagina does not, feeling rather soft and yielding. A frequent replacement of the protrusion, recumbent position for some time, injections of cold water, and finally the application of a sponge, three or four inches long and an inch thick, covered with linen, and changed for the sake of cleanliness twice a day, form the means generally sufficient to correct this complaint. The alternate use of *mercury* and *nux vom.*, every other evening a dose (six glob.) will be at the same time very beneficial, and should not be omitted. Complete conjugal abstinence during the cure is necessary.

LEUCORRHEA. *(Fluor Albus.)*

This disease, commonly called *whites*, is the most frequent, troublesome and weakening complaint of the female sex. It consists of a slimy, mucous discharge, variously colored, and of different consistency. It occurs usually between the age of puberty and the critical period, and is seldom seen later than this, except when discharges of a similar kind are excited in consequence of uterine disorganization. If it manifests itself in children, or even infants, it is either on account of a want of cleanliness, or local irritation, such as is produced by pin-worms, etc.

Weakly females, of a nervous, relaxed, or easily excited temperament, are more obnoxious to it; and the more refined or over-civilized our present state of society becomes, with its legion of pleasures, inactivity of body, idle and late hours, bad literature, and immoderate use of tea, coffee and spices of all kinds (we mention here only the increased use of vanilla,) the more easily will this disease be engendered.

The symptoms and exciting causes of leucorrhea are so various, that we have to divide it into several species. The most convenient division is that into an idiopathic and symptomatic; the former being of primary origin, not excited by other diseases, while the latter appears only in consequence of other diseases or by specific poisons, such as venereal, etc. We could subdivide these species into several classes, enumerating the peculiar characteristics of each class, but it would carry us too far in a treatise like this, whose aim is only practical usefulness; in this respect the above classification will be sufficient.

The primary or idiopathic fluor albus is mostly caused by constitutional predisposition in females, whose sexual organs are easily excited by menstruation, conjugal affinity, pregnancy and parturition. This is particularly the case, when stimuli of another kind are added, such as the use of highly seasoned food or exciting drinks, tea, coffee, vanilla-chocolate, etc.; or when powerful mental emotions, distress and languor depress vitality in such a degree as to prevent its lively re-action. Anything which accomplishes the latter, severe loss of blood, over-exertion, want of rest and sufficient nourishment, etc. will excite the primary fluor albus at once.

The secondary or symptomatic leucorrhea is caused by the presence of other diseases in the system, such as pin-worms, polypi, scirrhous or cancerous degeneration, dislocations and prolapsus of the uterus, piles, want of cleanliness, etc., and usually disappears as soon as these disorders are removed or cured.

The discharge itself differs as to quality, quantity and consistency; if it proceeds from the vagina alone, it resembles thick cream and is not ropy; while that from the uterus has the consistency of jelly, and adheres tenaciously wherever it lodges. The color varies; sometimes it is whitish, milky, yellow, even greenish; as to its acridity, it is mild or corroding; the latter quality predominates mostly in the chronic form.

The treatment of this disease is frequently very difficult, and should be conducted by a skillful physician. We can only mention a few remedies, useful to begin with, advising the reader to pay particular attention to the use of tepid water injections, or frequent sitz-baths of tepid water, for the sake of cleanliness as well as supporting the cure of the primary form of leucorrhea.

The treatment of the secondary form requires injections and sitz-baths of cool, even cold water, as the best means to restore the tone and strength of the relaxed parts involved in this disease;. the cure can also be promoted toward the end of the disease by the application of the wet bandage.

The following are the principal remedies in the commencement of the treatment.

Pulsatilla.—Discharge thick, like cream, sometimes creating an itching around the affected parts; in young girls before menstruation.

Sepia after the above remedy, if the discharge is yellowish, greenish, of a fetid odor and corroding; constipation with frequent bearing down sensation.

Cocculus.—Discharge of a reddish hue before and after menstruation, with colic and flatulency.

Calcarea carb.— Whitish, corrosive discharge, ren in children; in adult females of a lymphatic ، nstitution, light complexion, having copious and ،oo frequent menstruation, attended with diarrhea, itching and burning in the private parts.

Sulphur, in chronic cases of almost every kind, if the above remedies have failed to cure.

Take four doses of a remedy on four consecutive evenings, then await the result for a week; at the expiration of which, if not better, select another remedy and take it in the same manner. The diet must be nourishing, but not flatulent; avoid the exciting causes of this disease, particularly exposures to cold and damp, luxurious living and late hours.

DISEASES OF THE UTERUS.

The womb may be rightly considered the center of the female sexual organs, on account of its location as well as of the importance attached to its functional activity, although as to the latter, the ovaries may claim an equally distinguished position. A great variety of diseases can occur in the uterus, either naturally from mal-formation, or acquired from constitutional or other causes. In some women this organ has been found wanting entirely, in others of too small a size. Cases are on record where its opening, the so-called mouth of the uterus appeared closed, and others, where no internal cavity at all existed, and the whole uterus formed one compact mass; in a few instances the womb had increased in size throughout, or partially on one side only. All these abnormities, however, are very rarely found, and interest as such more the physician; they admit of but little remedial interference.

But the abnormal conditions as to the position of the womb are of greater interest and practical value, because they occur so much more frequently, and offer more or less in every case, a chance of cure by the proper medication.

PROLAPSUS UTERI. (*Falling of the Womb.*)

From the position the womb naturally occupies, hanging suspended on four ligaments in the middle of the pelvic cavity, the reader can easily infer that it can change its position in all directions; one of the most frequent consists in a sinking down, to a less or greater extent, into the lower part of the pelvic cavity, sometimes so far down as to protrude externally. An event of this kind must produce considerable derangement in the female organism. The most prominent symptoms of this disorder are fully described by Dr. Dewees, as follows: "The symptoms, characterizing this complaint will be modified by the greater or less descent of the uterus in the vagina : they will be intense in proportion to the extent of the displacement; but in all there will be a sense of something sinking in the vagina, as if the perineum were sustaining an unusual weight; with a dragging sensation about the hips and loins; a desire to make water, sometimes without the ability to do so; or if it do pass, it is reluctantly, and oftentimes painfully hot—a sense of faintness, and occasionally a number of nervous or hysterical feelings and alarms, which almost overwhelm the patient. A pressure and feeling about the rectum, resembling a slight tenesmus, sometimes importunely demand the patient's attention, which, if she obey, almost always end in unavailing efforts. The pain in the back is sometimes extremely distressing while the patient is on her feet, and gives to her walk the appearance of weakness in her lower extremities. A benumbing sensation shoots down the thighs, especially when the woman first rises upon her feet; or when she changes this position for a horizontal one. In some few instances, the woman is obliged to throw her body very much in advance; or is obliged to support herself by placing her hands upon her thighs when she attempts to walk. But all these unpleasant symptoms subside almost immediately if she indulge in a recumbent posture, and this circumstance pretty strongly designates the disease."

However well marked the above symptoms are, particularly the one last mentioned, they are not sufficiently so to prevent mistakes from being made in its diagnosis. An examination of the parts involved should never be omitted, as without it we never can be positive in pronouncing the disease prolapsus uteri. Dr. Dewees relates a case of this kind. "I was consulted by a lady, who had long suffered almost every symptom recorded above; I pronounced her disease to be a prolapsus of the uterus; and without an examination per vaginam, had a pessary made for its support; but, to my sad mortification, when I was about to apply it, a careful examination proved that no such condition existed, and that all the unpleasant symptoms had arisen from a thickening of the neck of the bladder."

But not all practitioners avow frankly, like Dr. Dewees, their errors in this respect; after once having pronounced a similar complaint to be falling of the womb, they rather persist in their opinion, even after having ascertained by actual examination that no prolapsus exists. In this manner falling of the womb has been multiplied in such a degree, that it may be counted now among the fashionable diseases, which any lady of standing or delicate feeling ought to have, at least a touch of it. While patients complain of symptoms, similar to prolapsus, some physicians hastily and without examination pronounce them to indicate this latter disease, thereby forcing their remedial action into a direction, often detrimental to the welfare of their patients. We have seen women tormented for years under a treatment against prolapsus uteri by several physicians, who, one after another, had readily yielded to the incorrect diagnosis of the preceding one. All that time these patients had not the slightest real symptoms of prolapsus or dislocation; they were afflicted with neuralgia, congestion, induration of the womb, etc., diseases which soon were removed by a rational (homœopathic) internal treatment.*

Rheumatism of the uterus is a frequent cause of feelings resembling falling of the womb, and a practitioner should be very cautious in pronouncing the existence of a disease so dissimilar in its treatment from the former.

Although a great many physicians yet adhere to the use of pessaries and abdominal supporters in the cure of prolapsus uteri, we must confess that we have never seen such good result as would induce us to persist in their use. Since we have become acquainted with the specific power of homœopathic medicines and the tonic virtues of cold water in the form of a sitz-bath and wet bandage, we have discarded the use of pessaries and other supporters, almost entirely.

---

* While writing the above, we see in an article on uterine displacements by Dr. Ramsay, in the Boston Medical Journal, similar complaints made as to the fictitious prevalence of prolapsus uteri; which, according to the statement of that gentleman, is far greater in the South than in other parts of the country. We can testify in some degree to the truth of this fact ourselves Residing in Cincinnati we had for a number of years frequent occasion to attend ladies from the South, the complaints of many of whom resembled greatly the symptoms attending prolapsus uteri. According to their statement they were suffering from prolapsus, having been told so repeatedly by their physicians; yet, upon examination, not the slightest trace of such a disease could be detected. We are glad to see that this professional error attracts the attention of those who best can correct it.

In the commencement of a cure for this disease, the patient should remain in a lying posture for a length of time, also otherwise refrain from active exercise as much as possible. The wet bandage, twice a day renewed, and frequent sitting-baths of short duration, will be of great benefit. We found the following remedies the most beneficial in this disease.

Belladonna and *sepia*, alternately, every other morning a dose (six glob.,) until better, at least for one week; during the next week the patient discontinues the medicine, but repeats the above prescription during the week following.

These medicines are succeeded, if necessary, after six weeks, by *calcarea carb.*, to be taken in the same manner.

Sometimes, during the above treatment, a dose

Hear Dr. Ramsay : "This *prolapsus* question has been a hobby for many a pretender to secure fame, and scores of women South have been injected per vaginam with sulph. zinc, nit. arg. *et id omne genus*, to their serious detriment, for the mal-position of an organ from which they never suffered. Any man, with a thimbleful of brains, who will put himself to the trouble to examine the anatomical situation of the womb, will see at a glance, that the organ, in its normal and physiological condition, is not easily prolapsed, at least not with the facility once supposed. We admit *real* prolapsus is too common ; but at the same time we protest against referring every little uneasy sensation in the hypogastric region to uterine descension. It is high time we were awakening from this unprofitable and unmeaning slumber, with regard to female affections, etc. God speed the time for the benefit of our *wives* and *daughters*."

of *nux vom., platina, opium, cocculus* and *ignatia* may be found necessary, if the patient exhibits a good deal of nervousness with constipation.

All stimulating diet is strictly prohibited; no coffee, no tea can be allowed; but good nourishing food is beneficial. Other displacements of the womb may occur in various directions and degrees.

The *retroversion* of the uterus, or that state wherein the womb is turned over backwards, occurs perhaps the most frequently, and produces many disturbances in the alvine and urinary discharges by actual pressure on the rectum and bladder, frequent hemorrhages from the womb, fluor albus and menstrual irregularities.

The anteversion of the uterus presents a deviation in a direction opposite to the former, the fundus uteri inclining towards, or even resting on the bladder, the neck and mouth of the womb towards the rectum. A distress similar to the former is the consequence.

In both cases, it becomes necessary to apply to a competent physician, who, after careful examination will replace the parts, thus removing the pains and sufferings. An operation of this kind is not in the least painful or exposing, and should be submitted to by the patient very readily as the only means to correct the evil at once.

During the course of these diseases follow the same hygienic rules as stated on preceding page.

INFLAMMATION OF THE WOMB.

Before puberty congestion and inflammation of the womb very rarely occur, and after this period the greatest liability for their appearance takes place during menstruation, pregnancy and childbed. At such times the womb is more predisposed to be affected by morbific causes, which at other times would scarcely reflect on it injuriously. The symptoms of an inflamed womb vary very much as regards locality and the circumstances which have produced them. The pains usually are burning, boring, stinging, pulsating, extending upwards into the abdomen and downwards to the thighs, with occasional intermissions. Accompanying these pains there is a sensation of heat and weight in the pelvic region, indicating mostly that part of the womb which is inflamed. The function of the uterus is more or less disturbed; either the menses have ceased or flow too readily; during pregnancy an early miscarriage, or during childbed a cessation of the lochia can be the consequence. Besides, there is present a greater or less degree of fever, preceded by chilliness, headache, and other concomitants of febrile re-action, even delirium.

In a disease of this kind the attendance of a physician is indispensable; yet the early application of the following remedies may prevent greatly the rapid progress of the disease, and promote its quick resolution.

Aconite and *belladonna*, dissolved in water, each one separately, in half a teacupful of water, alternately every two or three hours a teaspoonful, until better, will be generally sufficient to dispel the most severe symptoms.

In connection with the above a sitz-bath of ten minutes duration will be beneficial. The diet must be light, nothing but water-gruel and dry toast.

IRRITABLE UTERUS.
(Rheumatism and Neuralgia of the Womb.)

We have had occasion to speak of these complaints in connection with after-pains and other disorders of the lying-in. But they frequently appear idiopathic without being caused by, or complicated with any other disease. The most characteristic symptom of all of them is a pain, the seat and direction of which varies as it now proceeds from the small of the back and the lumbar region, now starts from the front part of the pelvis and radiates in all directions. The pain is mostly lancinating, boring, burning, tearing and heating. Dr. Dewees, who calls one species of this disease "irritable uterus," speaks of its local manifestation as follows: "Sometimes the patient represents the parts as being a little swelled; but this we believe is always transient. Walking, riding or indeed any kind of exertion, is sure to be accompanied or followed by severe lancinating pains within the pelvic cavity, especially from near, or in the course of, the urethra, to about the centre of the sacrum; and when the severity of the pain has abated, it subsides into a permanent dull pain in the same direction, but more diffused." The fact of the pains being increased by motion, erect position, and its abatement by a recumbent one, has given rise to an erroneous diagnosis, by mistaking this disease for falling of the womb. Yet a careful examination per vaginam, will convince us that, though the uterus is found to be lower in the vagina than common, it is not prolapsed or otherwise seriously displaced; but its neck is very sensible to the touch, somewhat shortened, because enlarged, and the mouth of the uterus more closed than in a natural state: the vagina itself feels hot and swollen.

The course of this disease is more chronic than acute, and involves more or less all the uterine functions. It appears mostly in paroxysms, brought on by certain exciting causes, such as erect position, over-exertion, conjugal connections, powerful mental emotions, even faults in diet; sometimes the most extreme point of the spine becomes so tender as to prevent the patients from sitting any length of time. To give the reader an idea about the variety of causes which may produce this disease, we insert here only a few of those which are related by Dr. Goosh. "In one patient it came on after an enormous walk during a menstrual period; in another, it was occasioned by the patient's going shooting with her husband not many days after an abortion; in a third it came on after standing for several hours many successive nights at concerts and parties; in a fourth, it originated in a journey in a rough carriage over the paved roads of France; in a fifth it was attributed either to cold or an astringent lotion, by which a profuse lochia was suddenly stopped, followed by intense pain in the uterus; in a sixth, it occurred soon after, and apparently in consequence of matrimony."

The treatment of these disorders should be conducted by a skillful physician. Before his attendance, an effort may be made to diminish the congestive tendency to the neck of the womb by giving *belladonna* and *sepia* alternately, every four days a dose (six glob.,) until better, in connection with tepid injections in the vagina, and tepid sitz-baths. During this time the patient should abstain from meat and other stimulating diet, keeping as quiet as possible.

POLYPUS OF THE UTERUS.

Polypi are indolent tumors, resembling fleshy or fungous tumors growing on the inner surfaces of cavities, for instance, the nose or uterus, where they root either with a broad basis or a narrow neck. They are sometimes very vascular and bleed readily on the slightest touch; for this reason their presence in the womb becomes very dangerous, particularly if they are of that kind which adhere, with a broad basis to its walls. As they grow, increasing in size and length, the uterus has to extend also, until it becomes stimulated to expel them, which is done in a manner similar to that during an abortion. The danger in such cases is very great, as inversion of the uterus may take place, which, in itself, constitutes a serious disorder.

Polypi of the uterus are met with as often among single as married women; frequent hemorrhages of the former, therefore, should lead us just as well to suspect the presence of a polypus of the womb, as it would do in the latter. The author observed a case of this kind in a girl of seventeen years, who, after having suffered for a considerable length of time, from oft repeated uterine hemorrhage, resisting the most varied treatment, was suspected by him to have a polypus of the uterus; upon examination its presence was verified, and the ligature soon restored the poor girl to health and happiness. If the polypus is very large, the ligature which the surgeon fastens around its neck, cannot at once cut its stem through; it has to be done gradually, which sometimes lasts a week or more. In one instance which came under my notice, the polypus, seven inches long by one and a half inches in diameter, was separated from its stem after ten days application of the ligature. During this time the profuse sloughing process required great cleanliness and absolute repose on the part of the patient, to prevent a slow fever which might have been created by the absorption of the sanious liquid and the hemorrhage in consequence of the extensive lesion. The most nourishing diet has to be chosen, such as oysters, poultry, eggs, soups of meat, etc.; if putrid fever sets in, acidulated drinks must be given. As regards remedies in this stage of the operation, *china*, *carbo. veg.*, *veratrum*, *arsenicum*, *lachesis*, are the most beneficial.

The application of the ligature is perhaps the quickest and best method to extirpate polypi wherever they may be located, if they only can be reached by it. Where they are thus not accessible, the Author has found the oft repeated application of a saturated tincture of *opium* on the polypus with a sponge, together with the internal use of *calcarea carb.*, third potency, every four days one dose (ten or twelve glob.,) the most successful method; he has thus cured polypi which could not be reached by the ligature. Although the process is a tedious one, yet it is after all gratifying to witness the removal of tumors which had resisted methods otherwise more expeditious, but wholly inapplicable.

It is unnecessary here for me to state that a disease of this kind requires the assistance of a skillful physician and the greatest willingness and patience of the sufferer while under his care.

SCIRRHUS AND CANCER OF THE WOMB.

To avoid the confusion and uncertainty which may arise in the mind of the reader by reciting the legion of premonitory symptoms appearing in advance of the above dreadful diseases, we will relate only the most important one which indicates the presence of the above disease more surely than any other symptom. A woman experiencing from time to time sudden *lancinating*, *darting* pains through the uterine region, should at once be on her guard and confide her case to an experienced physician. In most instances (this has been at least the author's experience,) in that early stage, the disease has not progressed far enough to be beyond the reach of judicious medication; the real open cancer has not at that time formed; there is, however, a cancerous tumor or induration in the neck of the womb, called scirrhus. This, if not resolved by proper medication, soon becomes a real cancerous ulcer, which in most cases terminates fatally.

The treatment of an open cancer must be left entirely to the directions of a physician, who, though not perhaps successful in curing the case, will be able, in most instances, of affording great relief.

In scirrhous affections of the womb, *belladonna* seems to be the most reliable, particularly if congestion to different parts of the system are present. *Conium* is generally the next remedy to be applied; after which *carbo veg.*, *sepia* and *sulphur* may be given in intervals of one or two weeks, if no amendment takes place.

Again let us entreat the reader to apply for medical aid as soon as the most important symptom which precedes these dangerous complaints, namely, the *sudden darting stitches* in the uterine region, make their appearance.

ULCERATION OF THE WOMB.

An ulcer is usually the result of previous congestion and inflammation, which, not having been resolved in the proper, manner excite indurations in the parts affected, in consequence of which suppuration sets in. The mouth and neck of the womb are more liable to this species of affection than any other part of the uterus. It is very difficult to distinguish this form of disease from cancer of the womb. But the absence of the lancinating pains not often found in an ulcerated uterus, but always present in cancer of the womb, makes the diagnosis sure. Generally but very few painful symptoms attend uterine ulcers, and their presence is frequently not detected except by an ocular examination through the speculum, which, in such cases, should always be instituted by the physician in charge of the patient.

There is always more or less discharge attending an ulceration of the womb; it varies in quantity, is mostly sanious and purulent in character, sometimes offensive and mixed with blood, thus differing materially from a common fluor albus.

The treatment of this disease is less difficult than other uterine disorders, the ulcers healing readily under the judicious treatment of a skillful physician.

In the beginning *belladonna* and *sepia* in alternation, every eight days a dose (six glob.,) are of great benefit to reduce the congestion and swelling of the parts affected. Afterwards the local application of *kreosot* and *calendula* becomes necessary, if the above remedies have not been successful. *Thuja* also, in injections, is highly recommended. Frequent injections of tepid water and the use of the hip bath, will accelerate the cure.

DROPSY OF THE WOMB.

At first sight it seems strange that an organ like the womb, having an aperture designed for the exit of menstrual blood, should be liable to secrete and retain water in such a quantity as to cause an enlargement sufficient to produce suspicion of pregnancy. Yet this often occurs, when, by some morbid process, the mouth of the uterus has been closed sufficiently to prevent the escape of the fluid, secreted within the womb.

Such a closure may be the result of accident, swelling, induration or adhesive inflammation, as is produced sometimes after a severe birth. The cause of the watery secretion itself in the cavity of the uterus, is not, as yet, well understood; the disease is said to be more frequent in women of a delicate, weak constitution, who have had many children; it also occurs more frequently under debilitating circumstances, after great fatigue, loss of blood, sedentary life, mental and bodily depressions.

The symptoms indicating dropsy of the womb present many of the signs of pregnancy, with this difference, that persons afflicted with it enlarge faster than during pregnancy, and that from time to time water escapes from the womb, either slowly or in gushes. This latter may also be the case during real pregnancy, and if so, the water had been collected within the membranes of the child, a form of dropsy already mentioned in the chapter on the disorders of pregnancy.

When the dropsy exists independent of a gravid uterus, the water may be let out by a catheter or a similar instrument; but even in that instance, though the patient be relieved at once, an internal treatment must follow, by which a renewal of the secretion is prevented. This is mostly accomplished by the exhibition of *arsenicum* and *sulphur*, every week two doses of each, for four weeks.

If not better *graphites* should be administered, every week one dose (six glob.,) for six weeks.

The counsel of a skillful physician should be had as soon as possible; although not absolutely dangerous, this disease being the result of constitutional ailments, requires skill and perseverance to overcome it successfully.

INFLAMMATION OF THE OVARIES.

In previous chapters, we have seen the important position which the ovaries occupy among the sexual organs; in them the first impulse starts for the development of all the other organs necessary to fœtal generation and growth. Consequently, a disease in these small bodies must be of the highest importance, and its cure should engage our full attention.

The ovaries are subject to many and serious derangements; one of the most frequent is inflammation, the result of congestion to these organs during sexual excitement, voluptuous imagination, disappointed hopes, etc., or caused directly by exposure to cold, suppression of the menses, external injuries, translation of rheumatism or gout from other places of the system, etc.

Its symptoms, although very characteristic, sometimes have been mistaken for inflammation of the womb. While the pain of the latter is more confined to the middle region of the abdomen, the pain in the former is found on one or the other side, just above the groins, where, on pressure, sometimes an enlargement may be discovered very painful to the touch. This is the ovarium swollen and tender, affected in a similar manner, as it is sometimes the case with the glands on the neck.

If the inflammation is an acute one, the fever present usually runs very high, connected with nausea, vomiting, strangury or a difficult voiding of scanty, high-colored urine; the thigh of the affected side generally feels numb, and the pains are increased by the patient suddenly rising up.

If the inflammation has assumed the chronic character, the above symptoms are present, but less severe. The fever, particularly, is at first very slight; the symptoms most obvious to the patient, are, perhaps, the stiffness and pain on moving the leg of the affected side, and a feeling of weight in the diseased region.

In either form, the inflammation of this organ requires immediate attention, as a neglect would produce the most disastrous consequences, such as suppuration, induration, watery effusions, etc., after diseases which may terminate fatally, and if not must at least produce sterility.

It is highly important to procure, forthwith, the proper medical aid.

Before this can be had, the following remedies may be given. *Aconite* in alternation with *belladonna*, every hour a dose (four glob.) This treatment frequently mitigates the symptoms in such a degree, that in a few hours all traces of the disease have disappeared.

The above remedies may be followed by *bryonia* if the disease is the result of a suppression of the menses; the same in alternation with *rhus* if cold or rheumatic disorders were the cause of it; in the latter case, give alternately every two hours a dose (four glob.)

Cantharides will be indicated when the disease is complicated with strangury or difficulty of voiding the urine.

Arsenicum, if the patient is very restless; should this remedy fail to relieve, the alternate use of *camphor* and *coffea* for a few hours, every fifteen minutes a dose (four glob.,) will quiet the most intense agony.

Conium, if the menses are suppressed, with severe semi-lateral headache, debility, blue circles around the eyes, variable appetite, nausea, swelling of the left ovary, with constant numb aching, contusive pain, and occasionally at night, paroxysms of severe sharp pains in the left side.

Tepid hip-baths will be of great assistance,

and if the inflammation, with fever, has risen to a very high degree, we may even be allowed to use the cold hip-bath, putting, at the same time, cold compresses on the head and covering well the other parts. The patient may remain in the bath from twenty-five to thirty minutes. After rubbing dry, a wet bandage should be applied around the parts affected, after which she should be put to bed, well covered. This treatment is particularly beneficial in that form complicated with retention of urine.

The diet should be light as in all inflammatory fevers; if thirsty, lemonade, water and crust-water are the best drinks.

OVARIAN DROPSY.

The ovaries are frequently liable to a degeneration of this kind. The water is contained within a sack or cyst, which has been formed by the enlargement of one of the ovarian vesicles. These tumors, filled with water, sometimes grow to an enormous size; their treatment must be left entirely to the physician, who should be in early attendance when it is yet possible to arrest their growth.

DISEASES OF THE BREASTS.

The breasts or mammary glands, forming a part, as it were, of the organs of generation, are involved more or less in all their changes during health or disease. This becomes particularly evident after delivery, when the breasts appear most prominently occupied in the secretion of that fluid which is indispensable for the maintenance of the offspring. It is during the performance of this important office, that the breasts more than at any other time, are liable to disease.

We have already, on page 280, mentioned the inflammation of these organs, known under the name of *ague in the breast*, and the swelling and suppuration which sometimes follows it. The reader will there find, also, the treatment laid down which we have found to act as the most beneficial during a large course of practice.

We will add, here, a few remarks about other morbid peculiarities of the breasts which may excite the interest of the reader. There have been observed several cases of a total deficiency of the breasts, occuring always in connection with atrophy of the ovaries and uterus. Sometimes we meet with very small breasts and with such as have diminished in size after they had been once developed; these are called atrophic breasts.

Still other instances are recorded where women possessed more than two breasts, or more than one nipple on each breast. In some cases the irregular breasts were located on the thorax, like the regular breasts; in others they occupied a place on the abdomen or in the groins.

Again, an hypertrophy or faulty enlargement of the breasts, is frequently observed, sometimes to such an extent as to border on a monstrosity; they have been found to measure forty-two inches in circumference, weighing twenty pounds.

Swellings of the breasts, different from permanent enlargement, frequently occur in consequence of lacteal disturbances and constitutional dyscrasies. Indurations of this kind have no malignant character.

It is quite different, however, with that hard tumor known as *scirrhus*, which is the forerunner of one of the most dreadful diseases, the *cancer of the breast*. We refer the reader to what we have said on page 315, about these diseases when located in the womb. The same remarks there, will apply to the scirrhus and cancer of the breasts. The scirrhus has been frequently cured when remedies were applied early; but an open cancer of the breast, as yet, belongs to the incurable diseases, although the palliation of its sufferings is within our reach.

CHAPTER III.

DISEASES OF NERVOUS FUNCTION.

There are but few diseases located in the nervous system, which belong, exclusively, to the female sex. The nervous complaints of the head, breast and other organs, such as nervous headache, spasmodic asthma, etc., which frequently can be traced to female peculiarities, do not constitute separate forms of disease, but being attributable to one and the same cause, are treated of together under one name.

HYSTERIA.

Our readers are perhaps aware that years ago hysteria or vapours (mostly known under this name in France) were quite fashionable. Ladies in high life at the European courts had nothing wherewith to kill the tedium of an otherwise taskless existence. To create variety and sustain intrigues, the hysterical fit, with its Protean character, offered the readiest and best means which, in the boudoir, saloon or promenade, could change, at once the scene from comedy to tragedy, and convert mirth into tears, wit into sympathy. This time has passed; the storms of revolutions have dispelled the sentimentalities of the age, and with them the supreme reign of hysterics and vapours in the female world.

At the present day, even the name has become obsolete and obnoxious to the fair sex. They dislike extremely to be told that they have the hysterics; they have nothing against it, however, to be nervous, which essentially means the same thing. We continue to use here, the name hysteria, because it is more significant of its origin.

The word hysteria is derived from the Greek, signifying the womb, because this disease was considered to be in close connection with the sexual peculiarities of the female. Upon the whole the correctness of this idea must be acknowledged, although different opinions are held, as yet, in regard to the primary cause of this complaint. Dr. Meigs gives his views as follows:

"The causes of hysteria are to be found in a great variety of conditions, both of internal and external origin; among them may be named a highly nervous and sanguine temperament, the pathological propensities of which are promoted by a sedentary life and luxurious living, hot rooms, hot beds, highly stimulating food, the use of wine, of aromatics; a mind ill regulated, indulged, intolerant of control, highly impressible. In such an individual, any abnormal degree of excitement that might serve to add to the purely physiological action of the reproductive organs, you could well deem sufficient to send its aura forth upon its mission of mischief throughout the entire economy."

Hysteria may be considered as a chronic disease, appearing from time to time in paroxysms. These latter are very irregular as to their characteristic symptoms, their intensity and exciting causes. Sometimes they occur at once without any apparent cause, or at least but a slight one, not proportionate to the effect produced, at other times exciting causes of the severest kind, must operate on the system for a length of time before a hysterical fit is produced.

To the unpracticed eye, the diagnosis of hysteria is rather difficult; but a careful observation of the pulse, the movements of which are not under the voluntary control of the patient, soon leads us to distinguish between severe forms of diseases and their hysterical counterfeits.

An hysterical attack usually commences by a feeling of tightness and fulness in the abdomen, which, rolling upwards like a ball, (*globus hystericus*,) produces, after it has reached the throat, a distressing sensation of choking and suffocation, followed by various efforts of the patient of relieving herself from the smothering and pressure, by eructations of wind, throwing around her arms, gasping for breath, etc; finally, pale and exhausted, she sinks down, unconscious, falling into spasms, apparently very violent, because the convulsive movements of the limbs and trunk are sometimes really frightful, and the spasms of the respiratory muscles in the highest degree alarming. But they soon cease; and the patient, having become tranquil, lies feeble and exhausted for some time, in a half sleepy state until she feels perfectly well again, or goes off into another spasm. If such paroxysms repeat often, one after the other, they resemble epileptic fits; in such cases, the history of the patient must guide us in our diagnosis. Moreover, in epilepsy the patients have all the symptoms of congestion to the brain, face red and bloated, features contorted, foam at the mouth, thumbs clasped in the palm of the hand, etc., symptoms which are seldom present in an hysterical fit.

A succession of these spasms may finally exhaust the whole system to such a degree that the patient sinks into a state of insensibility; the heart apparently ceases to beat, respiration becomes imperceptible and a general collapse of functional vitality takes place. As this state of apparent death sometimes has lasted for weeks without destroying the life of the patient, we should not cease to watch the patient, applying restoratives, until real signs of death have appeared or resuscitation takes place.

Hysteria, however, not always appears in the formidable form above described; it has various degrees of intensity, sometimes expressing itself only in crying, without a known cause, in unusual hilarity, laughter, excited conversation, etc. Dr. Meighs draws the following graphic picture of the versatile nature of hysteria:

"The hysterical woman, like the highly electrified thunder cloud, requires but the point to draw the flash. She sits, like Tam O'Shanter's wife,

> "Gathering her brows, like gathering storm,
> Nursing her wrath to keep it warm."

when, suddenly, and unexpectedly, some word, sign or gesture, gives the occasion; and we have reproaches, tears, screaming, laughter, sobs, wringing of hands, tearing of hair, clonic convulsions, tonic spasms, stertor, smiles like a May morning, loud laughter again, floods of tears and then a gradual return to a state of gentle composure, wherein the tenderest affections of the human heart come to resume, with unusual supremacy, their wonted sway over the soul."

During a fit of hysterical passion, it is all-imimportant, not to cross the patient in her expressions, or to heighten her distress by opposing arguments; it is of no use, but can do a great deal of harm. If she should fall into a spasm, she should be laid on a couch or bed, her clothes loosened on the parts where they may be tight around her, fresh cool air should be admitted into the room, and cold water sprinkled into her face; if the latter looks red and bloated, cold water may be poured on it from a height, or ice laid on the top of it, at the same time that her feet are put in a hot bath; if the face looks pale, this treatment is not necessary. Until the proper medicine can be procured, a camphor bottle may be held under her nose from time to time which frequently arrests the spasms at once.

As sudden mental emotions are the most frequent causes of hysterical fits the latter will be controlled by the remedies indicated for these various conditions of the mind. If caused by

Excessive joy: Coffea, opium.

Fright or fear: Opium, aconite. bellad., igna.

Anger, violent: Chamomile, bryonia, nux vom.

Anger, suppressed: Ignatia.

Grief: ignatia, phosphor. acid, staph., coloc.

Home Sickness: Phosphor. acid, mercury, capsicum, staph.

Unhappy Love: Hyoscyam., ignatia, phosphor. acid.

Jelousy: Hyoscyam., lachesis, nux vom.

Mortification and Insult: Bellad., ignat., platina, coloc., staph., puls.

Contradiction and Chagrin: Chamom., bryo., ignat., coloc., aconite, nux vom., platina, staph.

Indignation: Coloc., staphysagria.

Give of the medicine selected, four globules, dry on the tongue, every fifteen or thirty minutes, or dissolve twelve globules in half a teacupful of water and give a teaspoonful in the same intervals; after the attack, lengthen the intervals to three, six and eight hours, until the patient is relieved.

Of the above remedies *coffea, ignatia, pulsatilla,* and *aconite* will be the most suitable in almost any hysterical fit while it lasts. To prevent its return, however, medicines should be given which are able to remove the hysterical constitution. This can only be done by a careful treatment instituted by a skillful physician. Among the remedies for this purpose the following are the most important: *natrum mur., calcarea carb., sepia, sulphur.*

They may be taken in their order, each one for six weeks, every week a dose (six glob.)

During this time the patient should keep a strict diet, abstain from coffee, tea and all spicy substances, and if possible, should have recourse to all the strengthening appliances of the water-cure, such as frequent sponging-baths, sitting-baths, etc. If constipation is present, injections of cold water and the wet sheet around the abdomen; if inclined to congestive headache, cold foot-baths are recommended. The patient must take exercise in the open air after a cold bath, in order to promote the necessary re-action.

As already stated, hysterical paroxysms have, of late years, diminished in frequency and intensity; they, indeed, very seldom occur at present; and if they do, it is under circumstances, extraordinary and so severe that the attendance of a physician is at once required. As to the causes which have led to this singular phenomenon, the opinions of the writers vary. Dr. Dewees remarks: "It would be difficult to point out the causes of the diminution of this disease within the last thirty years in this city, though the fact is certain, so far at least as we can rely upon our own observations. Have the temperaments most liable to this disease been changed by either physical or moral causes? Certain it is, that at present we are rarely called upon to attend in an hysterical paroxysm, whereas formerly such calls were frequent." Perhaps the most potent agency having a tendency to diminish the hysterical diathesis, may be found in the fact that our present times exhibit less sickly sentimentality but more healthful activity and more sound utilitarian views, in all departments of life, than any former age.

INDEX.

Abortion, 234
Abnormal erotic sentiment, 200
Absence of " " 204
Acid stomach during pregnancy, 222
After-birth, 256
After-pains, 260
Ague in the breast, 280
Anteversion of the womb, 308
Apparent death of the infant, 254
Appetite, derangement of 215
Asphyxia of the infant, 254
Asthma, 228
Blood, spitting of 228
Breast, gathered 280
" ague in the 280
" abcess of the 282
" diseases of the 322
Cancer of the womb 315
Cessation of the menses, 199
Change of life, 199
Character of woman, physical, 21
Character of woman moral and intellectual, 31
Child's water, 251
Childbed fever, 268
Childbed, mania in 272
" melancholia in 272
Childhood, 61
Chloroform, use of 153
Chlorosis, 182
Colic, menstrual 197
Colic-pains during pregnancy, 223
Color, changed of milk, 286
Congestion during pregnancy, 206

Congestion of the head, 229
" of the liver, 227
" of the lungs, 228
Confinement, 261
Constipation during pregnancy, 221
Convulsions puerperal, 232
Cough, hacking 229
Cramps in legs, back and stomach, 223
Crusade, moral 159
Crying spasmodic 231
Delivery, 253
Depression of spirits during pregnancy, 230
Derangement of appetite, 215
Destiny of woman, 43
Deterioration of milk, 283
Diarrhea during pregnancy, 219
" preceding labor, 241
Diet during confinement, 263
Difficulty of swallowing 222
Diseases of women, 179
" of generative organs, 292
Diseases of sexual development, 181
Diseases of nervous function, 325
Diseases of the vagina, 297
" of the uterus, 302
Dislocation of the womb, 308
Displacement " " 308
Dropsy " " 317
Dyspepsia during pregnancy, 222
Dropsy, ovarian, 322
Dysury during pregnancy, 224
Eclampsia gravidarum, 232
Education, finished 103

INDEX.

Education, physical _____ 62
" intellectual ____ 73
Egg-beds, _____ 89
Elopement. _____100
Erotic sentiment, abnormal__200
" " absence of 204
Ether, use of _____153
Exercise, _____ 71
External parts, inflammation of _____293
External parts, wounds on __294
Fainting, _____230
Falling of the womb, _____303
Fashion, _____111
Feet, swelling of _____213
Fever, childbed _____268
" during pregnancy, __206
" milk _____279
Flooding, _____193–195
" after parturition, _____258 59
Fluor Albus, _____298
Fœtus, development of ____147
Generative organs, diseases of 292
Girl, _____ 53
Green sickness, _____182
Gymnastics, _____ 71
Hacking cough, _____229
Head-ache _____230
Head, congestion of the ___229
Heart-burn during pregnancy, _____222
Heart, palpitation of _____228
Hemorrhage during parturition, _____258
Hemorrhage during pregnancy _____209
Hemorrhoids during pregnancy, _____210
Hymen, imperforation of the 292
Hysteria, _____325
Icterus, _____225
Imperforation of the hymen 92
Incontinence of urine, ____225
Infancy, _____ 53
Infant, apparent death of the _____254
Infant schools, _____ 71
Inflammation of external parts, _____293

Inflammation of the ovaries 319
" of the womb __309
Intellectual faculties, _____ 37
Introduction, _____ 13
Irritability of the womb, ___310
Ischury during pregnancy, _224
Itching of the private parts, 295
Jaundice, _____225
Labia, oedematous swelling of the _____294
Labor, _____243
" natural, preternatural 244
" too sudden termination of _____245
Labor, protracted _____246
" sudden cessation of __247
" excessively painful __250
" commencement of ___242
Labor-pains, spurious or false, _____248
Laughter, spasmodic _____231
Leucorrhea, _____298
Limbs, swelling of _____213
Liver, congestion of _____227
Lochial discharge, _____264
Lochia, _____264
" supression of _____ 266
" excessive and protracted _____267
Lochia, offensive, sanious __267
Lungs, congestion of _____228
Maiden _____ 84
Maiden lady, _____120
Mania in childbed _____272
Marks, _____147
Marriage, second _____165
Married lady, _____132
Matron, _____169
Menses, cessation of the ___199
Menstrual colic, _____197
Menstruation, appearance of 84
" abnormal appearance of _____186
Menstruation tardy, _____187
" suppressed, ___190
" too copious __ 93
" too long duration of _____196
Menstruation too late and scanty, _____196

INDEX. 335

Menstruation, deviation of, 196
" too difficult, painful, 197
Menstruation, cessation of the, 199
Milk, deterioration of, 283
" changed color of, 286
" changed quality of, 286
" suppressed secretion of, 287
" excessive secretion of, 288
" deficiency of, 289
Milk fever, 279
Milk leg, 270
Miscarriage, 234
Monthly period, appearance of, 84
Moral and intellectual character of woman, 31
Moral sentiments, 34
Mother, duties of a, 154
Music, 107
Nausea during pregnancy, 217
Navel, pouting out of the, 151
Nervous function, diseases of, 325
Neuralgic pains, 230
Nervousness, 325
Neuralgia of the womb, 310
Nipples, soreness of, 290
Nursing, 273
" diet during, 278
Ovarian dropsy, 322
Ovaries, 89
" inflammation of, 319
Ovule, fecundated, 146
Pain in the right side, 227
Pains, neuralgic during pregnancy, 230
Palpitation of the heart, 228
Parturition, 240
Phlegmasia alba dolens, 270
Physical character of woman, 21
Piles during pregnancy, 211
Plethora during pregnancy, 206
Pleurisy, 228
Polypus of the womb, 312
Pregnancy, 206
" signs of, 144
Private parts itching of the, 295

Prolapsus of the womb, 303
" uteri 303
Pruritus, 295
Puberty, abnormal appearance of, 181
Puerperal convulsions, 232
Quality, changed, of milk, 286
Quickening, 145–150
Recapitulation, 175
Retroversion of the womb, 308
Rheumatism of the womb, 310
Right side, pain in the, 227
Sabbath schools, 128
Salivation during pregnancy, 214
Scirrhus of the womb, 315
Secretion, suppressed, of milk, 287
Secretion, excessive of milk, 288
Sexual development diseases of, 181
Show, 242
Signs of pregnancy, 144
Sleeplessness during pregnancy, 230
Sneezing, spasmodic, 231
Societies, benevolent, 127
Sore nipples, 290
S, asmodic, laughter, crying, sobbing, sneezing, yawning, 231
Spasmodic pains during pregnancy, 223
Spitting of blood, 228
Sterility, 205
Stomach, acidity of the, 222
Stroma, 89
Strangury during pregnancy, 224
Swallowing, difficulty of, 222
Swelling oedematous of the labia, 294
Swelling of the feet, 213
" " limbs, 213
Tooth-ache during pregnancy, 214
Ulceration of the womb, 316
Urination, painful difficult and interrupted, 224
Urination, involuntary, 225

Urine, incontinence of......295
Uterus, diseases of the......302
    " prolapsus of the......303
    " anteversion of the...308
    " retroversion of the...308
    " inflammation of the 309
    " irritable......310
    " Rheumatism of the.310
    " neuralgia of the......310
    " polypus of the......312
    " scirrhus of the......315
    " cancer of the......315
    " ulceration of the......316
    " dropsy of the......317
Vaccination,......61
Vagina, diseases of the......297
Vagina prolapsus of the......297
Varicose veins......212
Veins varicose......212
Vertigo......229
Vomiting during pregnancy.217
Waters the......251
Whites,......298
Widow,......161
Woman's physical character..21
    " moral and intellectual character,......31
Woman's destiny......43
Womb, see Uterus
Wounds on external parts,...294
Yawning, spasmodic......231
Young lady,......84

www.ingramcontent.com/pod-product-compliance
Lightning Source LLC
Chambersburg PA
CBHW021153230426
43667CB00006B/372